GUIDE TO HOME IMPROVEMENT COSTS

GUIDE TO HOME IMPROVEMENT COSTS

Bryan Spain

Routledge
Taylor & Francis Group

LONDON AND NEW YORK

First published by Butterworth-Heinemann

This edition published 2011 by Routledge
4 Park Square, Milton Park, Abingdon, Oxon OX14 4RN
605 Third Avenue, New York, NY 10017

Routledge is an imprint of the Taylor & Francis Group, an informa business

British Library Cataloguing in Publication Data
A catalogue record for this book is available from the British Library

Library of Congress Cataloguing in Publication Data
A catalogue record for this book is available from the Library of Congress

ISBN 13: 978-0-7506-5873-7 (pbk)

CONTENTS

Preface   vii

Introduction   ix

PART ONE

1   Using a contractor   **3**
When to use a contractor   3
Choosing a contractor   5
Obtaining estimates and quotations   6
Making payments   7
Variations and extras   8
Making a simple contract   9

PART TWO

2   Doing it yourself   **15**
Building regulations   15
Planning permission   17
Party walls   22
Ordering materials   22
Programming   26
Raising the finance   28
Grants   29
Home improvements – are they worth it?   30

PART THREE

3   Hours, materials and costs   **35**
Fireplaces   35
External walls and chimney pots   37
Roofing   38
Doors   42
Windows   61
Kitchen fittings   66

CONTENTS

Wall, floor and ceiling finishings 70
Plumbing 71
Glazing 79
Electrical work 83
Decorating and paperhanging 84
Paths and edgings 91
Fencing 94
Patios 95
Walling 98
Timber and damp treatment 99
Disposal of material 102

PART FOUR

4   Building your own home **107**
Why self-build? 107
Finding a plot 108
Finding the money 112
Professional help 113
Preparing a budget 115
Making a success of self-building 118

PART FIVE

5   Total project costs **123**
Traditional house extensions 123
PVC-U conservatories 125
Loft conversions 125
Swimming pools 126

PART SIX

6   Tool and equipment hire **129**

PART SEVEN

7   General construction data **135**
The metric system 135
Conversion equivalents (imperial/metric) 136
Conversion equivalents (metric/imperial) 136
Temperature equivalents 137
Areas and volumes 138
General building information 138

Glossary 145
Useful addresses 149
Index 151

PREFACE

The dramatic rise in house prices in recent years has meant that more home-owners are choosing to improve and enlarge their existing homes rather than move to another house! This has led to an upsurge both in DIY activities and in work for building contractors operating in the domestic construction market.

The public is overwhelmed with cost information on domestic appliances, holidays and cars but hardly anything on the cost of home improvements is available. This book has been written to fill that gap.

There has never been more interest in DIY activities than at the present – see the number of magazines and TV programmes devoted to the subject. Whilst building professionals – architects, surveyors, builders and engineers – have had access to information on building costs through price books for more than 150 years – there are hardly any sources of building cost data available to the general public.

This book is written for two types of home-owner. First, for those who are too infirm or too busy to carry out their own improvements and repairs. The information in the book will tell them what they should be paying a contractor for a wide range of items of work and should reduce the chances of overpaying for work. Second, for those who are capable of DIY work but wonder what savings in time and money could be made if they carried out the work themselves instead of using a contractor.

The contents of the book are in two main parts. First, general advice on how to employ, pay and manage contractors plus a guide to cowboy spotting – unfortunately they don't always wear Stetsons and spurs! Information on contracts, planning permissions, building regulations, finance and grants are also included.

Second, hundreds of home improvement items are set out in tabular form displaying times, material costs and skill levels for each activity. Other sections include information on building your own house, total project costs and mensuration formulae. A glossary of terms and a list of useful addressess complete the contents.

PREFACE

There are many women working in the building industry these days and where the pronoun 'he' appears in this book, it is intended to apply to both men and women. Although every care has been taken in the preparation of the book, neither the publishers nor I can accept any responsibility on the result of the use of the information it contains.

I would welcome constructive criticism on the contents and scope of the contents for use in future editions. I have received help from many individuals and firms in the preparation of the book and would like to thank the following:

- Jeld-Wen Ltd

- Woodfit Ltd

- Build Store

- *Build It* magazine.

I would particularly like to thank John Craggs for his help and support on matters affecting planning permissions and building regulations, Eddie Millership of Aurora 2000 for tweaking the software, Simon Young for technical support and Martyn Hocking, editor of *Build It*, for permission to use the magazine as a source for some of the information in Part Four.

Bryan Spain
January 2003

INTRODUCTION

The most important section of this book is contained in *Part Three: Hours, materials and costs*. The information displayed here is unique and will allow the reader to make a judgement on whether he is paying the right amount to building contractors and also whether it is worth sacrificing his spare time to complete a home improvement task instead of paying for the work to be done.

Domestic construction work can be placed under two main headings, repairs and improvements. Repairs are essential and must be undertaken to preserve the value of the property and the integrity of the structure. Improvements, however, are usually optional and are carried out to improve the quality of the home-owner's lifestyle.

In this section the information is set out in tabular form under the following headings:

- Description
- Quantity
- How long? hrs:mins
- Material costs
- Skill level
- Builders' charges.

DESCRIPTION

This column describes the work to be carried out and the items fall into two main categories:

(a) a complete operation usually involving only one trade or skill, e.g. repairing a broken roof tile;

(b) a cost per unit of an item, e.g. the cost per square metre of laying a concrete drive complete.

INTRODUCTION

Note that unless specifically stated, the cost of removing debris is not included in the rates. Item descriptions in the construction industry are usually derived from The Standard Method of Measurement but in this book the descriptions are set out using everyday words, but where technical terms are included, they are included in a glossary at the back of the book.

QUANTITY

The items measured are described by volume, area, length or enumerated and the following abbreviations are used:

no	number
m	linear metre
m^2	square metre
m^3	cubic metre.

HOW LONG? HRS:MINS

This is one of the key features of the book. The time necessary to carry out each operation is expressed in hours and minutes. Four hours twenty minutes, therefore, appears as 4:20.

These figures reflect the likely time necessary for an average DIY enthusiast to do the work. The word average is important in this context and should be taken as someone who has a basic knowledge of the use of tools coupled with an interest and enthusiasm for DIY work.

The word average should also be applied to the standard of finish achieved. A DIYer who is a perfectionist and looks for a mirror finish on paintwork will obviously take longer to carry out a task than someone who just wants the job completed to a reasonable standard of finish.

To summarise, the times represent the average time it should take the average person working in average conditions working to an average finish. If two people are involved, the times can be halved.

Time spent preparing the work, lifting carpets, moving furniture and clearing up at the end of each work session are not included. The times stated assume that the place of work can be reached easily and an extra 20 to 30 per cent should be added if it is necessary to work off ladders.

MATERIALS COSTS

The costs are based on the average prices of small quantities of materials available at local DIY supermarkets that are part of a national chain. Sales, end-of-season promotions and other factors can affect these figures but the regional differences in material costs are not so pronounced as they used to be.

If buying materials from smaller outlets, however, the costs may be significantly higher. Based on an average material cost index in this book of 100, the following adjustments should be made:

England	
East Anglia	86
East Midland	85
Inner London	110
North	88
North West	87
Outer London	102
South East	93
South West	88
Northern Ireland	81
Scotland	85
Wales	86

Readers in East Anglia, therefore, should reduce the material costs by 7 per cent but for most purposes, the costs included should suffice.

SKILL LEVEL

Each activity is awarded a 'skill factor' grade ranging from 1 to 10 indicating the skill level needed to carry out the work. For example, scraping off wallpaper is level 1 – the simplest of tasks – but plastering is rated at 8.

Generally speaking, tasks that are rated 8, 9 or 10 should only be attempted by experienced DIYers. Breaking out openings in load-bearing walls should always be carried out by professionals and the same rule applies to anything but simple electrical work.

BUILDERS' CHARGES

There are two types of costs affecting builders' prices and it is important to understand the difference between them. Construction costs are the result of adding together the total cost of labour, materials, plant, tools and overheads.

This figure, plus profit, is what a contractor should be charging for his services and that is the figure that appears under this heading.

But this figure can be influenced by other factors such as the builder's workload. If he is busy, he will quote you above the going rate, if he is slack, he won't.

The size of the job can also influence the quotation. For example, it may only take a plumber a matter of minutes to replace a washer, and should charge, say £4. But if he has travelled an hour each way to do the job, you can expect a much higher bill than that!

Most builders look at the time it would take to do small jobs in terms of half-day units and will charge accordingly. The benefit of competitive rates will only appear when a few days' work is involved.

There are three main groups that carry out home improvement work:

(a) small firms having a manned office and registered for VAT;

(b) a one-man firm working from home;

(c) a tradesman who works in the evenings and weekends outside his full-time job.

In this book it is assumed that the firm in (a) above will carry out the work although VAT has not been included in the costs unless stated.

The assessments of labour hours and material costs are based upon the author's knowledge and experience and it is unlikely that all of the figures quoted in this book will apply to every reader. The information should be accurate enough, however, to provide a valuable source of information to help home-owners in the upkeep and improvement of their properties.

PART ONE

USING A CONTRACTOR

When to use a contractor   3

Choosing a contractor   5

Obtaining estimates and quotations   6

Making payments   7

Variations and extras   8

Making a simple contract   9

1

USING A CONTRACTOR

WHEN TO USE A CONTRACTOR

The cost of having major improvements done to your house is in the same league as buying a car – something that needs to be done carefully and only after much thought and planning. When spending large sums of money, you will want the finished product to be of a high standard so, unless you are experienced at DIY work, you need to give some thought on whether your skills will be able to produce the standard of finish you want.

If you have no DIY skills at all, or if you do not have the time to undertake the work, then you must employ a contractor to do it for you. But if you are reasonably experienced and believe that what you lack in experience you can make up for in common sense and enthusiasm in some parts of the work, you should make a list of the parts of work you could carry out yourself. Take a single storey brick extension, for example.

Work to be done by the DIY enthusiast:

- excavation;
- concrete foundations and sub-floor;
- roof joists;
- fascia and soffit boards;
- rainwater pipes and gutters;
- floor tiling;
- kitchen fittings;
- glazing;

- painting;

- external paving.

Work to be done by contractor/tradesmen:

- brickwork and blockwork;

- plasterboarding;

- plastering;

- doors and windows;

- plumbing work;

- electrics;

- drainage.

Many of these items of work can be interchanged depending on the particular skills of the home-owner, but if enough thought is given to the allocation of work at the planning stage, working to a programme will be made much easier. In the above example, it would be unusual to find a contractor who is willing to carry out only part of the work. Self-employed tradesmen, however, would find the arrangement acceptable.

An important point to remember in planning a project is the likely effect it will have on the rest of the household. In the best situation, the property will be empty and in the worst, there will be young children in the house. Upgrading a bathroom or kitchen needs particular thought. Even if you have the skills to do the job but can only work at weekends and the work will take three weekends to complete, it sounds like a job for a contractor. After all, a family cannot operate normally without a bathroom or kitchen for three weeks or more.

If you did decide to build an extension in tandem with various tradesmen, you must make every effort to complete your sections of the work on time. Building even a simple structure such as a small extension still requires the work to be done in a certain order and each trade follows on in a prescribed sequence. Failing to meet a deadline would have a knock-on effect and tradesmen may not be able to fit in with your revised starting dates.

In the construction of a single-storey extension, there are two tasks that must be carried out by specialists. The first is the electrical work and although it may seem straightforward enough to fix a few plugs, there are stringent safety regulations to be observed, and the needs of insurance companies and possible future owners of the property need to be considered.

The second category of work is breaking out the opening between the existing building and the new extension. Unless the opening will be created within the width of an existing opening where the existing lintel will continue to support the overhead wall, this work must be carried out by a specialist.

Another factor to be considered in the decision on whether to use a contractor or do the work yourself is the question of cost. Can you afford to employ yourself – assuming, of course, that you are capable of doing the work? If you are self-employed or benefit from paid overtime, you may decide that it would be cheaper to employ a contractor.

CHOOSING A CONTRACTOR

Everyone has heard stories about builders being unreliable although most of them are apocryphal! Nevertheless, careful thought should be given to appointing a builder because he will become part of your household for a while, so you must make sure that you are happy with every aspect of this arrangement, not only the financial side of things.

The best way to engage a builder's services is by recommendation. Someone who has worked satisfactorily for a relation, friend, work colleague or neighbour is obviously the first choice.

The worst choice is using someone who knocks on your door. Reputable firms and tradesmen have no need to adopt this approach – their order books are usually filled with repeat orders from satisfied clients or from recommendations.

A typical line would be, 'Excuse me, but we have just finished a job round the corner and have some tarmac left over. I've just been looking at your drive and we could do it for you for about £400, that's a 75 per cent discount! We could start straightaway if you moved your car and have the whole thing finished in a couple of hours. So if you'll just move your car . . .'

First, they haven't just finished a job round the corner, unless it's another rip-off. Second, it will cost you more than £400 – that's just for openers – they'll find some extra work to charge you for. Third, it isn't a 75 per cent discount or anything like and it will probably cost you more to put right their shoddy work afterwards. Fourth, don't move your car!

Don't argue with these people. Just say no, but if they persist, say you'll have to ask your partner who works for the Police/Inland Revenue/Customs and Excise – any mention of these occupations and your doorstep will quickly become a cowboy-free zone!

If you cannot find a recommended contractor, look in *Yellow Pages* for a member of one of the following organisations:

Federation of Master Builders (0207–242 7583)
Institute of Plumbing (017108 472791)
Electrical Contractors Association (0207–229 1266)

Before accepting a quotation, always discuss the matter of insurances with the builder. He should have current cover for Employer's and Public Liability and All-Risks and ask to examine the policies. The Federation of Master Builders offer a MasterBond insurance that includes cover for the extra cost of employing another contractor if the first one goes bankrupt.

OBTAINING ESTIMATES AND QUOTATIONS

There is a general misconception that estimates and quotations are different but they are the same in the eyes of the law. But because this misconception is widespread, it pays to make sure that you and the builders have the same understanding on the documentation.

Some builders regard an estimate as an approximation of the value of the work and a quotation as a firm offer to carry it out. The quotation should be as detailed as possible because it can be a useful tool in the valuation of any variations that may occur.

The quality of the quotation is directly related to the quality of the enquiry document. If the enquiry was '...just build me a kitchen extension...' you shouldn't be surprised if the quotations vary by as much as 100 per cent.

Providing drawings and a specification should produce more accurate quotations but for small parcels of work, it is probably sufficient to present the builders with a list of the work you want carrying out and talk them through it. Whichever way you decide to do it, make sure that all of the firms bidding are given the same information otherwise their offers will not be comparable.

Ideally, you should invite three or four firms to tender but don't place a firm on the list that you are not happy about because, as night follows day, it will make the lowest bid! When the bids come in, you should examine them carefully to check that that the offers cover the scope of the work you want carrying out.

You may not decide to accept the lowest offer for a number of reasons but when you have made a choice, invite the contractor to your house and run through the work list again to avoid any misunderstandings. For example, sort out who is responsible for moving the furniture and clearing away the rubbish.

Discuss working and access times. The question of toilet arrangements and brewing-up facilities all need to be settled before work commences. Unless the work is small enough to warrant only one payment on completion, you will need to agree to stage payments.

Finally, if you are happy to go ahead, accept the offer in writing (although a verbal acceptance is enough) including all of the points that were agreed at the meeting.

MAKING PAYMENTS

Most disputes that occur on small building works are related to money and payments so it is important that there is a clear understanding with the contractor before the work starts on how much and when the payments are to be made. There are three main ways of paying for building work: up front, stage payments or on completion.

Payments up front are not recommended! It is hard to imagine any circumstances where money should be paid out before the work commences. Some contractors or tradesmen may ask for an advance payment but it should always be refused.

If a firm is so financially insecure that it can't afford to fund the early stages of a project, it should not be in business. It would be extremely difficult to persuade a struggling firm to return to the job later on to carry out some remedial work!

Perhaps an advance payment could be made to a tradesman operating a one-man-business who is well known to you and completely trustworthy. But even then you should not make a payment by cash or cheque. Most requests for advance payments are based upon the need to buy materials to start the job so, if you decided to help, open an account at a local merchants and give the tradesman the authority to order materials on the account.

Place a ceiling on the account to match the value of the materials needed to carry out the work. The benefit of this arrangement is that if things do go wrong at least you would be the legal owner of the materials.

Paying by stages is the normal method when the value of the work is over say, £4000. But there is a right and a wrong way to pay for work this way. The wrong way is to make the payments on a time basis. For example, if a house extension is worth £12 000 and the period for construction is three months so it is agreed that you will make three payments of £4000 on the last day of each of the three months.

Due to the builder's inefficiency, bad weather and the late arrival of materials, the work is only half completed after month two. So the value of work done to date is £6000 but £8000 has been paid producing an overpayment of £2000.

If things continue at the same pace only two thirds of the work will be completed by the end of month three, the value of work will be £8000 but the payments will be £12 000, with an overpayment of £4000. The situation now is that the builder has been paid in full but there is still a third of the work to complete! Can you imagine the energy the builder is going to put into finishing the job knowing that there is no more money to come!

Never arrange stage payments based on time. Never divide the contract sum by the contract period and pay the resultant figure monthly. Never enter into any arrangement that does not relate payments to progress.

Here is the right way to pay by stages. The work should be broken down into clearly defined sections and values placed against each one. Here is a typical breakdown for a house extension worth £12 000.

	%	£
Foundations	10	1 200
First floor joists	15	1 800
Wall plate level	15	1 800
Roofed in	20	2 400
Plastered out	15	1 800
Completion	25	3 000
	100	12 000

On a job of this size, it is worth considering holding back some retention money of say, 5 per cent, from payments including the final one. This is to ensure that any defects that appear after the work is completed will be attended to. The retention could be released three months after practical completion has been reached.

Most domestic improvement work is too small in value to warrant stage payments and the work is normally paid for on completion. Depending on the size and status of the contractor, I would expect jobs less than £5000 in value to come into this category.

VARIATIONS AND EXTRAS

Disputes over the payment for variations are common in the domestic construction market. It is not always how much should be paid, but whether any payment should be made at all. You should appreciate that builders generally do not like variations to the work they originally contracted to do because they disrupt and delay the programme. Here are examples of three different types of variations. First, you want your front door renewed and accept a quotation based on the type and size you have chosen. Just before the work is complete, you tell the builder you want gold-plated door handles instead of plastic.

You have changed the specification so you must pay for it. You may even have to pay more than the difference between the cost of plastic and gold-plated handles if the contractor has to make a special journey to change the handles and particularly if the merchant won't take them back.

Second, if you have ordered gold-plated handles in the first place but the builder fixes plastic ones, it is obviously his error. He must either offer you a reduction for the difference in cost between the two or comply with your original specification.

The third type of variation is more complex. It involves the builder having to carry out work different from that originally envisaged and the argument is based on whether an experienced contractor should have foreseen the problem and allowed for it in his quotation.

For example, you want a couple of extra radiators fitted and you accept the quotation. When the contractor comes to do the work he finds that the floor under the fitted carpet is solid but he allowed in his quotation for laying the pipework between softwood joists. He asks for extra money for cutting chases in the concrete floor. Should he be paid?

Depending upon other circumstances, probably not. His experience should have told him to inspect the floor before he quoted. But what if he was replacing a bath and found that the joists under the old bath were rotten and needed renewing? Experienced or not, the builder could not have reasonably foreseen this and he should be paid for the extra work.

Some of the variations that occur are not as black and white as the three listed above and some sensible compromise must be reached. Try and avoid taking up an entrenched position because, if the dispute turns serious and lawyers become involved, guess who will be the winners!

The valuation of variations can also be a source of disputes but a detailed quotation can help to reduce this problem. But there are bound to be occasions when the extra work cannot be valued by referring to the quotation or any other documents.

The only possible solution then is to ask the builder for a quotation for the extra work before it is carried out. You will be able to decide whether to accept the quotation or not, beforehand. Unless, you have complete faith in the contractor, insist on knowing the cost of extra work in advance. Don't accept '... we'll sort the money out at the end ...' because that approach usually ends in tears. You don't have to pay until the work is complete, of course, but you must know the likely cost in advance.

MAKING A SIMPLE CONTRACT

Three things are needed to make a contract. An offer, an acceptance and consideration. Consideration is the benefit (usually money) that the person making the offer will receive for carrying out his part of the contract. In home improvement terms, a quotation from a builder to build a conservatory would be the offer, approval to go ahead would be the acceptance and the price quoted to do the work would be the consideration.

The contract is in place when the acceptance is made – the consideration does not need to have taken place, only the intention to do so. Although most contracts are in a written format, a verbal form is just as binding but much more difficult to prove if a dispute went to the courts. An offer can be withdrawn at any time before the acceptance is made but the offer cannot be revoked afterwards. Acceptance is deemed to have taken place on posting but the withdrawal of an offer is not effective until it has been received.

If a counter offer is made the first offer becomes invalid. For example, you receive a quotation to build a conservatory for £8000 but think it is too high. You tell the builder that he can go ahead with the work for £7500 but the builder rejects it. You approach other contractors but receive quotes of £9000 and £10 000 for the work.

You then approach the first contractor and tell him that you will accept his offer of £8000 after all but he is not obliged to accept it because you made a counter offer of £7500. Any offer is deemed to be valid for a reasonable period of time. Lawyers can afford exotic holidays on the proceeds of defining the meaning of the word 'reasonable'!

If a party breaks the terms of the contract, the other party is entitled to receive damages for any incurred losses caused by the event. The damages are intended to restore the offended party to the position he was in before the terms of the contract were broken. They are not intended to punish the contract breaker.

Assessing the value of the damages can be complicated because it may involve matters of inconvenience and mental suffering but common sense normally prevails and settlements are usually reached without recourse to the courts. If you have been let down by a builder – slow working or poor workmanship, for example – your case could be strengthened by asking the builder to put the matter right himself.

When things are moving towards a dispute, keep notes of all the conversations and telephone calls. Photographs should also be taken if relevant to the dispute. Claims for less than £5000 can be taken to the Small Claims Court without the need to employ a solicitor. Although many solicitors offer a 'no-win no-fee' arrangement these days, their deductible expenses and fees can reduce any damages awarded to a level far below the sum awarded.

If you decide to use the Small Claims Court service, obtain a leaflet EX301 (and others) from a Citizens' Advice Bureau. Quite often the threat of taking someone to court will have the desired effect. Here is a typical letter that you could send.

2 July 2003

Dear Sir

I wrote to you on 17 May and 3 June about the leak in my conservatory roof that appeared two weeks after you completed your work but you have not replied to either letter. In the letter dated 3 June I advised you that unless you carried out the repair within 14 days I would have the work carried out by another firm and send you the bill.

Again, you didn't reply so I have had the work done. I enclose a copy of the bill, £247.43 plus VAT, and I expect to receive your cheque for this amount within the next 7 days. Failing this, I will issue a county court claim against you.

Yours faithfully

Before making a claim, you should make sure that the person you are claiming from has the funds to pay if you win the case. You may have to pay a fee to start your claim and this is dependent upon the amount you are claiming and there may be further fees if the case is defended although these costs are added to the sum you are claiming if you win your case. If you are receiving income support you can apply to be exempt from paying these fees.

The hearings are generally low key affairs and the judge usually tries to create an informal atmosphere to put those people who are not used to court hearings at their ease. You may appeal if your claim is unsuccessful although it must be based on proper grounds and not just because you believe the judge made the wrong decision.

PART TWO

DOING IT YOURSELF

Building regulations 15

Planning permission 17

Party walls 22

Ordering materials 22

Programming 26

Raising the finance 28

Grants 29

Home improvements – are they worth it? 30

2

DOING IT YOURSELF

BUILDING REGULATIONS

Approval for building regulations is required for almost all building works including:

- house extensions;
- internal alterations to a house;
- change of use to living accommodation;
- removal of load-bearing walls;
- new bathrooms, showers, toilets or drains;
- loft conversions;
- cavity wall insulation;
- new chimney or flues;
- underpinning;
- new windows in walls or roof; and
- re-roofing with roof coverings that are substantially heavier or lighter than the original roof covering.

Not all new buildings are subject to building regulations approval and these include greenhouses, some porches and conservatories. For porches and conservatories to be exempt they must comply with the following conditions:

- built at ground level;
- single storey;
- floor area of less than 30 square metres;
- there is a door separating the new building from the main house;
- the roof and walls of a conservatory are mainly translucent; and
- the glazing complies with Part N of the regulations.

This glazing requirement is comparatively new and is intended to protect people from serious injury if the glass is broken. The glass is required to break safely, i.e. in small pieces or be in small panes or be protected.

Although the above exemptions to porches and conservatories mean that building regulations are not required, there is still an obligation on the owner or the builder to ensure that the new building does not cause any danger to health or safety or affect any escape routes required under the fire safety regulations.

Building regulation approval is required for the construction of some garages and carports. All attached garages and single-storey detached garages with a floor area over 30 square metres require approval.

Detached garages with an area of less than 30 square metres that are constructed mainly of non-combustible material or more than one metre away from the nearest boundary do not require building regulations approval. Similarly, carports that are open on two sides and are less than 30 square metres floor area do not require approval. Converting a garage into living accommodation requires approval.

Several important changes were made to building regulations on 1 April 2002. Among the changes were new requirements for glazing and insulation and advice on these changes should be obtained from your local Building Control offices.

The Building Control service must be notified in advance if any work is to be carried out requiring approval. The work will be inspected during construction and a Completion Certificate issued on satisfactory completion. This certificate should be retained by the householder to show any future owner of the property that the work was approved.

Although it is not mandatory to have plans drawn up, it is usually worth-while for other reasons such as tendering purposes. If you are in any doubt, you should consult a professional or contact your local Building Control offices for advice.

To apply for approval you need to send in either a Full Plans submission or a Building Notice submission. You will need professional advice on which is appropriate for the work intended to be carried out.

For a Full Plans submission, you will have to pay a Plan Charge on submission and an Inspection Charge after the first inspection. For a Building Notice submission, you will be required to pay the Plan Charge and the Inspection Charge on submission.

The fees for building regulation approval vary in different local authorities and the figures set out in Table 1 exclude VAT and should be regarded as indicative only.

TABLE 1

| | Full plans submission | | |
	Plan charge £	Inspection charge £	Building notice submission £
Garage or carport up to 40 square metres	30	90	120
Garage or carport 40 to 60 square metres	90	160	250
Extension up to 10 square metres	220	included	350
Extension 10 to 40 square metres	100	250	500

The building regulation fees for loft conversions are usually regarded as the same as those for extensions with a minimum fee of £327.66 because of the complexity of this type of work.

PLANNING PERMISSION

Planners often receive a bad press but this is because we only hear about them when something goes wrong. Most of the responsibility for planning matters has been handed down from central government to local planning authorities operating within local government. Planners carry out their duties for the good of the whole community and, even a cursory inspection of some of the housing development that took place in the 1930s, will show what an important role they play today.

There are various appeals procedures in place if you are unhappy with any planning decision that affects you and, if they fail, you can always pursue the matter through the courts. The ideal situation, and it exists in many parts of the United Kingdom, is where planning authorities use their powers to protect the character of the area whilst allowing individuals sufficient freedom to alter and improve their own property.

If you have queries on planning matters you should telephone or call at your local planning office for information or advice. You will probably be surprised by the courtesy and help that you will be offered. Planners, however, cannot change the law, and although the Chief Planning Officer does have some discretionary powers, he usually acts for the common good when applying them.

Here are examples of when you will need planning permission:

- alterations to a flat or maisonette affecting the external appearance;
- dividing your house to create a separate home;
- building a new house in your garden;
- dividing part of your house for commercial purposes;
- making changes that contravene the original planning permission.

You do not need to apply for planning permission where work is of a minor nature coming under the heading of 'Permitted Development Rights.' These rights vary in different parts of the country and are more strict in conservation areas, national parks, areas of outstanding natural beauty and other designated areas.

Here are the main types of work that may require planning permission:

- extensions, conservatories, loft conversions, dormer windows, verandahs, enclosing existing balconies and roof alterations;
- new buildings on your land such as garages, swimming pools, garden sheds, greenhouses and storage tanks;
- porches;
- fences, walls and gates;
- patios, hard standings, paths and driveways.

Extensions, conservatories, loft conversions, dormer windows, verandahs, enclosing existing balconies and roof alterations

You need planning permission if:

- you intend to build an extension between your existing house and a highway unless there will be 20 metres or more between the new extension and the highway. The term 'highway' includes footpaths, roads, bridleways and byways;

- the area covered by the extensions exceeds more than half of the area of land surrounding the existing house;

- the extension is higher than the highest point of the original house;

- any part of the extension is more than 4 metres high and within 2 metres of any boundary;

- in the case of a terrace house (or any house in a conservation area, national park or area of outstanding natural beauty), the volume of the original house would be increased 10 per cent or 50 cubic metres, whichever is the greater;

- for any other house outside those restricted areas, the volume would be increased by 15 per cent or 70 cubic metres, whichever is the greater;

- in any case, the volume of the original house would be increased by 115 cubic metres or more (the volume is calculated using the external measurements of the original building and includes outbuildings and garages if within 5 metres of the main building).

You do not normally need planning permission to re-roof your house or insert skylights or roof lights but if you live in a conservation area, national park or area of outstanding natural beauty, you will need permission to undertake work that will change the shape of your roof.

In other areas, you will need permission only in the following cases:

1. The new work would be higher than the existing roof.

2. The dormer would extend beyond the plane of any existing roof slope facing a highway.

3. The roof extension would increase the volume of a terraced house by more than 40 cubic metres or 50 cubic metres for any other type of house.

New buildings on your land such as garages, swimming pools, garden sheds, greenhouses and storage tanks

Most buildings under this heading do not require planning permission but there are some exceptions. These include:

- erecting a building that would be nearer to the highway than the nearest part of the existing building unless there would be 20 metres or more between the highway and the new building;

- the area covered by the new building exceeds more than half of the area of land surrounding the existing house;
- the new building is going to be used for commercial purposes including parking a van or storing materials;
- the new building will be more than 3 metres high or 4 metres high if it has a ridged roof (measured from the highest adjacent land);
- in a listed building, the proposed new building is more than 10 cubic metres;
- in a conservation area, national park or area of outstanding natural beauty, the proposed new building is more than 10 cubic metres.

You will need planning permission to erect a storage tank in the following circumstances:

- the capacity of the tank exceeds 3500 litres or is more than 3 metres above ground level;
- the tank is for storing heating oil and would be nearer to the highway than the nearest part of the existing building unless there would be 20 metres or more between the highway and the new building;
- the tank would be used for the storage of Liquefied Petroleum Gas (LPG) or any other liquid fuel excepting oil.

Porches

You will need planning permission to erect a porch if:

- the external ground area of the porch exceeds 3 square metres;
- it is more than 3 metres high above ground level;
- it would be less than 2 metres away from a boundary of a dwelling house with a highway.

Fences, walls and gates

You will need to apply for planning permission to erect or make alterations to a fence, wall or gate if you live in a listed building or if the height of the fence wall or gate exceeds 1 metre next to a highway used by vehicles or over 2 metres elsewhere.

You do not need planning permission to plant hedges or trees in most situations but there are exceptions such as maintaining sight lines. Check with your local planning office if in doubt. If you live in a conservation area, you do not need permission to remove a fence, wall or gate.

Patios, hard standings, paths and driveways

Building hard surfaces around your house does not require planning permission unless significant works of embankments or terracing are involved or the surface is to be used for the parking of commercial vehicles.

You should consult your local highways department if you are proposing to build a drive crossing an existing path or verge and you may need planning permission if the drive provides access on to certain types of road.

Planning permission generally

For most home improvement work, planning permission is not required although changing the external appearance of your house in a conservation area, national park or area of outstanding natural beauty could be an exception.

If you are uncertain about your position, contact your local planning office and ask their advice. Explain your problem and they will advise you on the best way forward – this could save you money by taking the necessary action before incurring the expense of appointing someone to draw up plans for you.

After you send in the application forms and pay the fees, your application will be acknowledged and placed on the Planning Register so that it can be inspected by members of the public. They will also notify your neighbours in certain circumstances and may place a notice in the local newspapers.

Your application will either be handled by a planning officer in the department or will be sent to the planning committee for consideration. In either case, your application will be judged on its merits. You should receive a decision within eight weeks but you will be notified if it will take longer.

If your application is refused or is subject to conditions, you should talk to members of the planning office to seek the best way forward. You are entitled to a second application without charge if you apply within twelve months of the decision made on the first application.

If you think that the decision is unreasonable, you are entitled to appeal to the Secretary of State for Transport, Local Government and the Regions. This appeal must be made within six months of the council's decision and you can also appeal if the council do not decide on your application within eight weeks.

If you wish to know about the appeals procedure you should write to the Planning Inspectorate, Customer Support Unit, Room 3/15, Eagle Wing, Temple Quay House, 2 The Square, Bristol BS1 6PN and ask for two free booklets, 'Making Your Planning Appeal' and 'Guide to taking part in planning appeals'.

Making an appeal should be treated as a last resort because it can take several months to reach a decision. It is far better to maintain good relations with your local council and to try and solve any problems in consultation with them.

The fee for planning permission for alterations and/or extensions to a single dwelling house including garages, fences, walls is £110 in all parts of UK. VAT does not apply to planning permission fees.

PARTY WALLS

If you intend to carry out work that is covered by one of the following categories you must find out whether the work comes under The Party Wall Act 1996:

- work on a shared wall or structure shared with another property;
- building a free-standing wall or a wall of a building up to or astride the boundary with a neighbouring property;
- excavating near a neighbouring building.

Dealing with party wall matters can be quite complicated and you may need professional advice. Start by studying a booklet called *The Party Wall Act 1996* from your local planning office. It is free and published by the Department of the Environment, Transport and the Regions.

ORDERING MATERIALS

There has been an increase in the number of DIY outlets in recent years and the competition between them is fierce. DIY enthusiasts are now able to buy materials at a level that was previously only available to trade customers.

Most of these stores will deliver materials to your home or you can hire one of their vans on an hourly basis. Although ordering most materials is fairly straightforward, there are some items that require a little more thought.

Concrete is an example. Unless you go to a ready mix concrete firm, you cannot buy a cubic metre of concrete, but only the ingredients. Details of how much to order are given under the heading of *Concrete work*. Another factor that must be considered when ordering is making an allowance for waste because it would be unusual if you were able to place the exact amount of material that you had ordered.

Excavation work

Excavated material bulks after it has been dug out, i.e. it occupies a greater volume after excavation than it did in the ground. It would be a mistake to order a skip sized six cubic metres if the pre-excavated volume of earth was also six cubic metres. The following figures represent the average amount of bulking that takes places for various types of earth.

	%
Sand and gravel	10–15
Clay	20
Hard materials, rock	40–50

Concrete work – site mixing

Although most people's perception of concrete is that of a basic building product, it is a highly complex material and research is constantly being undertaken to find out more about this versatile product. In domestic building work the two most commonly-used mixes are 1:2:4 and 1:3:6.

The numbers refer to the proportion by volume of cement, sand and aggregate used in the mixes. The 1:2:4 mix is usually placed in paths, floors and walls whilst 1:3:6 is predominately used in foundations. These ingredients are usually sold by weight and the following table gives the amount of material needed to mix one cubic metre of concrete.

	1:2:4 mix kg	1:3:6 mix kg
Cement	240	170
Sand	520	550
Aggregate	950	1 000

Take a simple example. You intend to lay a foundation to garden wall and the dimensions are 9000 mm long by 300 mm wide and 150 mm deep. Multiplying these figures produces a volume of 0.41 cubic metres. The foundation would be 1:3:6 mix so you would need to order the following quantities of material to complete the work. Note that these figures include a 5 per cent allowance for waste.

Cement
$170\,kg \times 0.41\,m^3 + 5\% = 73\,kg$

Sand
$550\,kg \times 0.41\,m^3 + 5\% = 237\,kg$

Aggregate
$1\,000\,kg \times 0.41\,m^3 + 5\% = 430\,kg$

It is unlikely that you will be able to buy the exact quantities you need but the above figures should help to reduce any unnecessary waste. Sand is sometimes sold by volume instead of weight so you may need to adjust the above figures accordingly – there are 1600 kg to a cubic metre of sand.

Concrete – ready mixed

For larger jobs, such as laying a concrete drive, it is probably better to have the concrete delivered ready mixed. This is a very competitive market, so shop round for quotations. Concrete is usually delivered in four, five or six cubic metre loads and you will pay extra per cubic metre for delivering 'air' – this is the term used in the industry for part loads.

Some firms specialise in small orders and they usually advertise their services in *Yellow Pages*. It is worth preparing some part of your garden, such as an uncompleted path or a base for a shed, to receive any concrete left over.

Brickwork

The thickness of brick walling is usually described as half brick thick (102.5 mm) or one brick thick (215 mm thick). There are many different ways of arranging laid bricks, called the bond, but in most domestic building situations a stretcher bond is the most common.

There are 62 bricks per square metre in a half brick wall and 124 in a one brick wall. These figures include a 5 per cent allowance for waste. The question of ordering materials for mortar is a little more complicated but the following figures should help.

Brickwork is usually laid in cement or gauged mortar. Cement mortar is normally used in foundation walls for extra strength and consists of one part of cement to three parts of sand (1:3). Gauged mortar is used in most other work and is made up of one part of cement, one part of lime and six parts of sand (1:1:6).

Here are the dry weights of materials per cubic metre of mixed material.

	Cement mortar kg	Gauged mortar kg
Cement	440	230
Lime	–	120
Sand	1820	1920

The average quantity of mortar needed in the construction of one square metre of a half brick thick wall is 0.03 m³ and 0.06 m³ for a one brick thick wall. If you are only building a small amount of brickwork, it may be better to buy the mortar already dry mixed but it is much cheaper to mix it yourself.

Floor and wall tiling

The following table shows the number of tiles you will need for each square metre of surface to be covered. You should add 7.5 per cent to these figures and more if the areas are interrupted by projections and the like.

Size of tile	Number of tiles
100×100 mm	100
150×75 mm	89
150×150 mm	44
200×100 mm	33
200×200 mm	25
225×225 mm	20
250×125 mm	32
250×250 mm	16
300×150 mm	22
300×300 mm	11
450×450 mm	5
600×600 mm	3
600×450 mm	4

When buying the materials make sure that you can obtain a refund on any unopened packs in case you have over-ordered. If you are fixing ceramic tiling, you will probably need a paste-type adhesive and this is sold in tubs of different sizes to cover between one and six square metres of tiling.

Painting

The covering capacity for paints depends to a large extent on the surface being painted, the type of the paint and the skill of the painter. The figures set out below are based upon brush application and are for square metres per litre of paint.

Surface	Undercoat	Gloss	Emulsion
Plaster	11–14	11–14	12–15
Brickwork	6–8	7–9	7–9
Blockwork	6–8	6–8	6–8
Woodwork	10–12	10–12	12–14

Wallpapering

The number of rolls of wallpaper needed to decorate a room will depend upon size of the pattern repeat, the number of doors and windows and the height of the room. The table below gives the number of rolls required for different-sized rooms with a standard number of doors and windows.

Perimeter	Height of papered wall						
m	2.10 m	2.25 m	2.40 m	2.55 m	2.70 m	2.85 m	3.15 m
8.50	4	4	4	4	4	5	5
11.00	5	5	5	5	6	6	6
13.50	6	6	6	7	7	7	8
16.00	7	7	7	8	8	9	9
18.00	8	8	8	9	9	10	10
20.50	9	9	9	10	10	11	12
23.00	9	9	10	11	12	12	13
25.50	10	11	12	12	13	14	14
28.00	11	12	13	13	14	15	16
30.50	12	13	14	15	15	16	17

If in any doubt about the number of rolls to order, order an extra one because nearly all retailers will allow you a refund on an unopened roll.

PROGRAMMING

The need for forward planning in any type of construction project cannot be stressed enough. The more thought put into organising the sequence of operations, even on a small house extension, the more smoothly and efficiently will the project run.

At its simplest, it would be foolish to waste time appointing a painter before you had someone in place to dig the foundations. Unless you are going to appoint a contractor to carry out all the work for you, you will have to examine the project and identify packages of work that you intend letting out to different trades. Bear in mind that some trades, usually Carpenters and Joiners, Plumbers and Electricians, will need to carry out their work in two visits for first and second fixing.

The Table 2 is a simple bar chart incorporating a list of activities relevant to a house extension and sequences of working. Once the extension is watertight, most of these activities overlap each other so that two or three trades can be working at the same time. Closely linked to the timing of the work is the need to order the materials and arrange for its delivery at the right time.

TABLE 2

Activity	1	2	3	4	5	6	7	8	9	10	11	12	13	14	15	16
Foundations	▬	▬														
Brickwork			▬	▬	▬											
Roofing						▬	▬									
Carpentry 1st fix							▬	▬								
Electrics 1st fix							▬	▬								
Plumbing 1st fix							▬	▬								
Plastering										▬	▬	▬				
Joinery 2nd fix												▬	▬	▬		
Electrics 2nd fix													▬			
Plumbing 2nd fix														▬	▬	
Decorating															▬	▬
Week number	1	2	3	4	5	6	7	8	9	10	11	12	13	14	15	16

Ideally, the programme should be linked to a schedule of stage payments so that the financial management of the job can be handled efficiently. When appointing contractors or tradesmen, the dates on your programme should be explained to them in the hope that they will arrive on time, do the work and leave on time.

In large projects a critical path analysis is prepared to identify those activities whose performance is critical to the job's progress. In smaller projects, such as house extensions, this is not necessary because almost every activity is critical and each trade is dependent upon the others to keep the operation running smoothly.

In the early stages of the building of a house extension before the structure is watertight, the weather can disrupt progress and this can affect the whole programme. It is wise therefore to build some float time into the programme to allow for any unforeseen delays. Adding an extra week at the end of the foundations before the external brickwork starts would be a sensible move in case the job gets off to a bad start.

If any delays do occur, it is important that the trades following on are kept informed as much as possible of any likely changes to their starting dates.

RAISING THE FINANCE

If you intend to have major improvement works done to your property, you will have to determine at an early stage how you are going to pay for them. If you have savings, it would seem that the most straightforward course would be to use them to finance the work.

Before doing so, find out the rate of interest you are receiving on the investment and compare it with the cost of borrowing the money. In normal circumstances, the cost of borrowing should always be higher than the return on investment but it is worth checking whether this is the fact in your case.

Hardly a post delivery goes by without some firm offering to lend money. Banks, building societies and finance houses appear to have cash mountains that they are determined to share with the general public. Sifting through these offers needs care because some of them, usually the ones with the lowest rates of interest on offer, only apply for a limited period. Sometimes, this can be as little as six months and then the rate reverts to the market rate or higher.

If you have to borrow money, the best place to start is with your own bank or building society. The days of these organisations being aloof and unapproachable are gone, so treat them as you would if you were making any other purchase. If you were buying a new car it is unlikely that you would deal with the first garage you contacted.

So ask your own bank for a quotation on the sum you want to borrow and see if you can better it elsewhere. If the difference is marginal, you probably better stick with your own bank where you are known. If a problem develops with the repayments, you may find that you will receive a more sympathetic hearing from someone who knows you.

GRANTS

There are several grants available to property owners and a series of government leaflets are available. These are set out in general terms in *'House Renovation Grants'* published by the Department of Transport, Local Government and the Regions and available at your local housing office or Citizens' Advice Bureau.

The following grants are covered:

- House Renovation Grant;
- Common Parts Grant;
- HMO Grant;
- Disabled Facilities Grant;
- Home Repair Assistance;
- Group Repair Schemes;
- Relocation Grant.

This book deals only with the House Renovation Grant. The type of work covered is:

- *making a building fit* – where the building falls below the level of fitness required by the local council;
- *to put a dwelling into reasonable repair* – this can also cover replacement work including damp courses, electrical wiring and guttering and repair work to roofs, timbers, walls' foundations, floors, staircases and plasterwork;
- *home insulation* – including loft, pipe and cavity wall insulation among others;
- *heating* – providing central heating or other facilities;
- *providing internal satisfactory arrangements* – covering work such as low doorways and steep staircases.

To qualify for the grant you must be in one of the following categories:

- *freeholder*;
- *leaseholder* with at least five years remaining on the lease;

- *landlord* who is either a freeholder or leaseholder with at least five years remaining on the lease;

- *tenants* with responsibility under the lease for carrying out repairs.

Grants are not available for holiday or second homes. The grant is means-tested to ensure that those most in need of help receive the benefit. The test takes into account the joint incomes of both partners and any savings over £6000. The grant is awarded for difference between the proposed cost of the work and the amount you are, deemed by the council, able to pay.

You will almost certainly require professional advice in determining the cost of the work and in making the application but these fees can be included in the cost of the works. The grant will be awarded provided that you complete the work within twelve months of the grant being approved, and that you use an approved contractor. Any breach of these and other conditions will require that the grant is paid back in full.

If you would like to know more about renovation grants you should contact your local council or Citizens' Advice Bureau.

HOME IMPROVEMENTS – ARE THEY WORTH IT?

A distinction must be made between improvements and repairs. Repairs, such as pointing, re-wiring, damp coursing and roof repairs, must always be carried out because neglecting them will cause further damage and reduce the value of the property.

Improvements – unless carried out carelessly and in poor taste – will generally enhance the property's worth. In general terms, you should only carry out improvements to your home to improve the quality of your own life. Evidence shows that spending £20 000 on a swimming pool will not increase the value of your house by £20 000 – it may even have a negative effect because some prospective buyers may not want the hassle of looking after a pool.

House prices have risen dramatically in recent years so it is sometimes difficult to assess whether the increased value is due to the extra demand for property or the intrinsic value of any improvements that have been carried out. With certain exceptions – they can be identified from the table below – it would be unwise to spend money on home improvements if you intend to move within two years.

One of the reasons why people move house is often because they need an extra bedroom. If you lived in a road where the average house price is £120 000 and you were faced with the choice of moving or building another bedroom on your house, you would face a difficult decision.

Spending £20 000 on a bedroom would increase the value of your house but owning the most expensive house in the road can be a drawback when it comes to selling it. Prospective buyers seem to judge property values by the general tone of the area as much as the quality of any individual house.

In this particular case, depending upon family needs, it would probably be better to build an extension and worry about future house values when the need arises. In planning to improve your house, it should be remembered that it is important to maintain a balance in the different types of room uses.

For example, building a second bathroom in a two-bedroom house is extravagant unless, of course, you intend to use the amenity yourself in the long term. Similarly, constructing a fourth bedroom in a house with only one sitting room would be unwise in normal circumstances.

The following table shows the value of construction costs likely to be recovered if the property is sold within two years of the work being carried out.

Type of work	%
Central heating	50–75
Garage	50–75
Double glazing	40–50
Loft conversion	20–40
Basement conversion	20–40
Conservatory	15–25
Bedroom extension	10–20
Kitchen extension	10–20
Porch	10–15

PART THREE

HOURS, MATERIALS AND COSTS

Fireplaces	35
External walls and chimney pots	37
Roofing	38
Doors	42
Windows	61
Kitchen fittings	66
Wall, floor and ceiling finishings	70
Plumbing	71
Glazing	79
Electrical work	83
Decorating and paperhanging	84
Paths and edgings	91
Fencing	94
Patios	95
Walling	98
Timber and damp treatment	99
Disposal of material	102

3

HOURS, MATERIALS AND COSTS

This part of the book contains data on the time necessary to carry out a wide variety of home improvements together with the material costs involved and the charges likely to be made by a contractor. This information is either set out as a composite item, complete in itself or as individual items that must be added together to produce the final figure.

FIREPLACES

Most DIY enthusiasts have the skill to take out and fix a new fireplace – the hard part is having the muscle to clear away the debris! Unless the existing concrete hearth can be broken up and bagged indoors, this is a job for two people.

These figures exclude the cost of removing the debris to a tip (see *Disposal of Material* on page 102) and the scrap value of the old fireplace.

				Skill Factor	
Take out timber fireplace surround size					
1050 × 900 mm	1 no	1:00	–	3	10.00
1200 × 1050 mm	1 no	1:10	–	3	12.00
1500 × 1200 mm	1 no	1:20	–	3	15.00

				Skill Factor	
Take out tiled concrete fireplace surround size					
1050×900 mm	1 no	2:00	–	3	15.00
1200×1050 mm	1 no	2:30	–	3	20.00
1500×1200 mm	1 no	3:00	–	3	30.00
Take out tiled concrete or stone hearth size					
1050×450 mm	1 no	1:20	–	3	12.00
1050×600 mm	1 no	1:30	–	3	18.00
1200×600 mm	1 no	1:40	–	3	24.00
Brick up opening after removal of fireplace, patch plaster walls, fix new softwood skirting 150 mm high, for opening size					
450×600 mm	1 no	8:00	15.00	5	115.00
600×600 mm	1 no	8:00	20.00	5	125.00
750×600 mm	1 no	8:00	25.00	5	135.00

The following figures represent hours and costs involved in fixing new fireplaces. There is a wide variation in the basic costs of fireplaces so the figures shown cover a range of costs.

				Skill Factor	
Supply and fix new fireplace after removal of existing (see above) including patch plastering around edges					
basic price, £120	1 no	6:00	125.00	4	200.00
basic price, £150	1 no	6:00	155.00	4	230.00
basic price, £200	1 no	6:00	205.00	4	280.00
basic price, £250	1 no	7:00	255.00	4	350.00
basic price, £500	1 no	8:00	505.00	4	600.00

Remember that if your existing chimney is used to extract smoke from an open fire and you intend to place a gas fire in the new fireplace, you should take expert advice on lining the flue. Although this can be done without the need for structural work, it is not a job for most DIYers.

EXTERNAL WALLS AND CHIMNEY POTS

This section deals with minor repair work to external brick walls and the replacement of chimney pots. Most repair work to brickwork involves the raking out and re-pointing of the mortar joints. In older buildings, the mortar sometimes decays and this is usually caused by driving rain or the penetration of water from a leaking gutter that has not been repaired.

Re-pointing decaying mortar is essential to maintain the integrity of a building and it can also enhance the appearance. The figures quoted are based upon the work being carried out at ground level and 20 per cent should be added for the cost of working off platforms and ladders.

				Skill Factor	
Rake out joints of brick walls and point up in mortar	1 m²	1:20	0.75	5	12.00
Rake out joints of chimney stacks and point up in mortar	1 m²	1:30	0.75	5	12.00
Cut out single brick from external wall and replace with new brick bedded in cement mortar	1 no	1:00	0.50	5	8.00
Cut out air brick from external wall and replace with new brick bedded in cement mortar, size					
215×65 mm	1 no	1:00	2.20	5	14.00
215×140 mm	1 no	1:10	3.00	5	16.00
215×215 mm	1 no	1:20	7.50	5	20.00

Sometimes in building improvements the cost of carrying out the work bears little relation to the value of the job itself. Replacing chimney pots is a case in point. The largest element of cost in this type of work is gaining access to the chimney stack.

Access can be gained by scaffolding, ladders or special platforms, each carrying its own cost. Because of this variation, the figures below apply to the time and cost of replacing the pots only, not the cost of gaining access.

				Skill Factor	
Remove existing chimney pot, hack away flaunching and set new terra cotta pot in cement mortar					
size 185 mm diameter 300 mm high	1 no	2:30	25.00	5	60.00
size 185 mm diameter 450 mm high	1 no	3:00	35.00	5	75.00
size 185 mm diameter 600 mm high	1 no	3:30	50.00	5	90.00
size 185 mm diameter 900 mm high	1 no	4:00	75.00	5	125.00

Dampness in external walls is often caused by existing flashings becoming detached from the brickwork. This is caused by the mortar becoming degraded and the flashing pulling away from the face of the wall. Rainwater then enters behind the flashing and causes unsightly staining to the internal walls and this can lead to more serious problems.

				Skill Factor	
Rake out joint in brick wall, re-fix flashing and point up in mortar	1 m	0:45	0.50	5	8.00

ROOFING

There is a wide variation in skill needed to carry out different types of roof repairs. Sometimes, the skill level is quite low for a simple task but gaining access to do the work may be difficult or daunting and hence it is advisable that these jobs are left to the experts.

This section is presented in three parts; repairs to pitched roofs, repairs to flat roofs and new work. Neither the cost of removing debris nor the hire of access equipment is included in the following costs.

Bear in mind, that where the cost is shown under Builder's Charges for replacing a single roof slate, it is only for the work and does not include any travelling time.

Repairs to pitched roofs

				Skill Factor	
Take off roof coverings for disposal					
tiles	1 m²	0:40	–	3	5.00
slates	1 m²	0:40	–	3	5.00
timber boarding	1 m²	0:50	–	3	6.00
battens	1 m²	0:20	–	3	3.00
underfelt	1 m²	0:05	–	3	1.00
flat sheeting	1 m²	0:15	–	3	2.00
Take off roof coverings and lay aside for re-use					
tiles	1 m²	0:50	–	3	6.00
slates	1 m²	0:50	–	3	6.00
timber boarding	1 m²	0:60	–	3	7.00
flat sheeting	1 m²	0:25	–	3	3.00
Take off roof tiles or slates for disposal					
ridge tiles	1 m	0:15	–	3	2.00
hip tiles	1 m	0:15	–	3	2.00
Take off roof tiles or slates and lay aside					
ridge tiles	1 m	0:15	–	3	2.00
hip tiles	1 m	0:15	–	3	2.00
Take up roof coverings from flat roof for disposal					
bituminous felt	1 m²	0:30	–	3	4.00
metal sheeting	1 m²	0:25	–	3	3.00
wood wool slabs	1 m²	0:30	–	3	4.00
firrings	1 m²	0:20	–	3	2.00
Take up roof coverings from pitched roof, lay aside for re-use					
tiles	1 m²	0:50	–	3	6.00
slates	1 m²	0:50	–	3	6.00
metal sheeting	1 m²	0:35	–	3	4.00
flat sheeting	1 m²	0:35	–	3	4.00
Inspect roof battens, re-fix loose and replace with new, size 38 × 25 mm					
50% of area					
250 mm centres	1 m²	0:40	1.60	5	5.00
450 mm centres	1 m²	0:35	1.40	5	4.00
600 mm centres	1 m²	0:25	1.00	5	3.00

		⏱	£	Skill Factor	£
100% of area					
250 mm centres	1 m²	0:40	3.20	5	10.00
450 mm centres	1 m²	0:40	2.80	5	8.00
600 mm centres	1 m²	0:40	2.00	5	6.00
Take off single slipped slate and re-fix	1 no	1:00	–	6	6.00
Remove single broken slate, renew with new Welsh blue slate	1 no	1:00	4.00	6	10.00
Remove slates in area approximately 1 m² and replace with Welsh blue slates previously laid aside	1 m²	3:30	0.00	6	28.00
Remove single slipped tile and re-fix	1 no	1:00	–	6	6.00
Remove single broken tile and renew					
265 × 165 mm	1 no	1:00	1.00	6	15.00
380 × 230 mm	1 no	1:00	1.40	6	16.00
413 × 330 mm	1 no	1:00	1.60	6	17.00
Remove tiles in area approximately 1 m² and replace with tiles previously laid aside					
265 × 165 mm	1 m²	3:00	–	6	38.00
380 × 230 mm	1 m²	2:40	–	6	26.00
413 × 330 mm	1 m²	2:20	–	6	24.00
Take off defective ridge or hip capping and re-fix including pointing in mortar	1 m	1:00	0.50	6	10.00
Take off defective ridge or hip capping and replace with new including pointing in mortar	1 m	1:00	3.50	6	18.00

Repairs to flat roofs

The lifespan of an average flat roof is between 25 and 30 years, although this can be extended by careful maintenance. Only minor patching repairs to bituminous felt roofing should be carried out by DIYers and more extensive work should be left to the experts.

Further, work to flat roofs involving copper, aluminium or lead is highly specialised and professional advice should always be sought for this type of work.

			Skill Factor	
Cut out crack in bituminous roofing felt, apply bituminous compound and apply proprietary sealing strip 150 mm wide	1 m	1:20	3.50	5	10.00
Cut out blister in bituminous roofing felt, apply bituminous compound and apply proprietary sealing strip to area approximately 300×300 mm	1 no	1:00	2.50	5	8.00

New roofing work

Laying roof tiles and slates is usually a job for an expert but the range of work being undertaken by DIYers seems to be increasing every year so this type of work is included below in this edition. The extra work involved at the verges, hips and eaves has been included in the overall rate.

There are two main types of roof coverings, slates and tiles. The most common slates (but not a common price!) are Welsh Blue and they are becoming more popular these days. An alternative is reconstituted slates made from slate dust. They are much cheaper but have the same general appearance as standard slates.

The two main types of tiles are concrete and clay. Again they are similar in appearance but the concrete tiles tend to lose their colour after a few years although this problem is not as bad as it used to be.

Specialist costs are shown for bituminous felt roofing – this is not a job for the amateur!

			Skill Factor	
Welsh Blue slates laid on underfelt and battens	1 m²	2:00	48.00	6	76.00
Ridge tiles	1 m	0:40	14.00	6	22.00
Reconstituted slates laid on underfelt and battens					
Marley Monarch	1 m²	1:40	24.00	6	46.00
Redland Cambrian	1 m²	1:40	28.00	6	50.00
Ridge tiles	1 m	0:40	8.00	6	16.00

				Skill Factor	
Concrete roof tiles laid on underfelt and battens					
Marley Plain	1 m²	2:00	20.00	6	36.00
Marley Ludlow Plus	1 m²	1:40	13.00	6	17.00
Marley Modern	1 m²	1:20	12.00	6	18.00
Ridge tiles	1 m	0:40	10.00	6	18.00
Clay roof tiles laid on underfelt and battens					
size 265×165 mm	1 m²	1:20	24.00	6	42.00
Ridge tiles	1 m	0:40	12.00	6	20.00

	Unit	Builder's charges £
Bituminous roofing felt laid flat		
two layer	1 m²	12.00
three layer	1 m²	18.00
Bituminous roofing felt laid sloping		
two layer	1 m²	13.00
three layer	1 m²	19.00
Layer of chippings bedded in hot bitumen	1 m²	3.00
Bituminous roofing felt skirting 150 mm high turned into groove		
two layer	1 m	8.00
three layer	1 m	11.00
Bituminous roofing felt in lining to gutter 400 mm girth turned into groove		
two layer	1 m	10.00
three layer	1 m	13.00

DOORS

There has never been such a choice of doors available than there is today. Fitting a new door is a comparatively cheap way of changing the appearance of a room and it can all be done in a reasonably short space of time. Doors can be placed in two main categories, internal and external.

Internal doors can be further divided into two types, solid and hollow. Solid doors can be quite expensive although the prices appear to be stabilising or even reducing in recent years. Hollow doors are made up of two layers of boarding filled in with an egg-box type material but with a timber support to the edges and middle rail.

Both solid and hollow doors may be covered in hardwood veneers that enhance the appearance (and the cost!) of the door. Examples of times and costs are set out below for the removal of existing doors and frames.

			£	Skill Factor	
Take off door from frame	1 no	0:20	–	3	3.00
Remove door frame and architraves	1 no	0:25	–	3	3.50
Take off existing ironmongery from door					
hinges	1 pr	0:15	–	3	2.00
bolt	1 no	0:10	–	3	1.50
deadlock	1 no	0:15	–	3	2.00
mortice lock	1 no	0:15	–	3	2.00
cylinder lock	1 no	0:15	–	3	2.00
push plate	1 no	0:10	–	3	1.50
handles	1 no	0:10	–	3	1.50
door closer	1 no	0:20	–	3	2.50

Most of the information on doors in this section is based on information supplied by Jeld-Wen UK Ltd, Doncaster (01302 394000).

			£	Skill Factor	
Internal flush hardboard doors					
Flush door 35 mm thick with hardwood lippings to vertical edges, hardboard faced, unprimed, size					
1981 × 381 mm	1 no	1:30	32.50	6	46.00
1981 × 457 mm	1 no	1:30	32.50	6	46.00
1981 × 533 mm	1 no	1:30	32.50	6	46.00
1981 × 610 mm	1 no	1:30	32.50	6	46.00

		🕐	£	Skill Factor	⚱
1981 × 686 mm	1 no	1:30	32.50	6	46.00
1981 × 711 mm	1 no	1:30	33.50	6	47.00
1981 × 762 mm	1 no	1:30	33.50	6	47.00
1981 × 838 mm	1 no	1:30	35.00	6	50.00
2032 × 813 mm	1 no	1:30	35.00	6	50.00

Flush door 35 mm thick with hardwood lippings to vertical edges, hardboard faced, primed size

1981 × 381 mm	1 no	1:30	42.50	6	56.00
1981 × 457 mm	1 no	1:30	42.50	6	56.00
1981 × 533 mm	1 no	1:30	42.50	6	56.00
1981 × 610 mm	1 no	1:30	42.50	6	56.00
1981 × 686 mm	1 no	1:30	42.50	6	56.00
1981 × 711 mm	1 no	1:30	43.50	6	58.00
1981 × 762 mm	1 no	1:30	43.50	6	58.00
1981 × 838 mm	1 no	1:30	45.00	6	60.00
2032 × 813 mm	1 no	1:30	45.00	6	60.00

Flush door 35 mm thick unlipped to vertical edges, hardboard faced, primed both faces, size

1981 × 381 mm	1 no	1:30	30.00	6	44.00
1981 × 457 mm	1 no	1:30	30.00	6	44.00
1981 × 533 mm	1 no	1:30	30.00	6	44.00
1981 × 610 mm	1 no	1:30	30.00	6	44.00
1981 × 686 mm	1 no	1:30	30.00	6	44.00
1981 × 711 mm	1 no	1:30	31.00	6	45.00
1981 × 762 mm	1 no	1:30	31.00	6	45.00
1981 × 838 mm	1 no	1:30	35.00	6	50.00
2032 × 813 mm	1 no	1:30	35.00	6	50.00

Internal flush veneered doors

Flush door 35 mm thick African Maple veneered finish, hardwood lipped, size

1981 × 457 mm	1 no	1:30	77.50	6	92.00
1981 × 533 mm	1 no	1:30	77.50	6	92.00
1981 × 610 mm	1 no	1:30	77.50	6	92.00
1981 × 686 mm	1 no	1:30	78.00	6	92.00
1981 × 762 mm	1 no	1:30	78.50	6	92.00

			£	Skill Factor	♦
1981 × 838 mm	1 no	1:30	81.00	6	95.00
2032 × 864 mm	1 no	1:30	93.00	6	97.00
Flush door 35 mm thick Sapele veneered finish, hardwood lipped, size					
1981 × 381 mm	1 no	1:30	55.50	6	60.00
1981 × 457 mm	1 no	1:30	55.50	6	60.00
1981 × 533 mm	1 no	1:30	55.50	6	60.00
1981 × 610 mm	1 no	1:30	55.50	6	60.00
1981 × 686 mm	1 no	1:30	55.00	6	60.00
1981 × 711 mm	1 no	1:30	55.00	6	60.00
1981 × 762 mm	1 no	1:30	55.00	6	60.00
1981 × 838 mm	1 no	1:30	58.00	6	72.00
Flush door 35 mm thick Beech veneered finish, hardwood lipped, size					
1981 × 610 mm	1 no	1:30	128.50	6	142.00
1981 × 686 mm	1 no	1:30	128.50	6	142.10
1981 × 762 mm	1 no	1:30	128.50	6	142.00
1981 × 838 mm	1 no	1:30	134.00	6	148.00
Flush door 35 mm thick White Oak veneered finish, hardwood lipped, size					
1981 × 610 mm	1 no	1:30	131.50	6	145.00
1981 × 686 mm	1 no	1:30	131.50	6	145.00
1981 × 762 mm	1 no	1:30	131.50	6	145.00
1981 × 838 mm	1 no	1:30	137.50	6	152.00
Flush door 35 mm thick Cherry veneered finish, hardwood lipped, size					
1981 × 610 mm	1 no	1:30	135.50	6	150.00
1981 × 686 mm	1 no	1:30	135.50	6	150.00
1981 × 762 mm	1 no	1:30	136.50	6	150.00
1981 × 838 mm	1 no	1:30	141.50	6	155.00
Flush door 35 mm thick Ash veneered finish, hardwood lipped, size					
1981 × 610 mm	1 no	1:30	119.50	6	133.50
1981 × 686 mm	1 no	1:30	119.50	6	133.50
1981 × 762 mm	1 no	1:30	120.00	6	134.00
1981 × 838 mm	1 no	1:30	133.00	6	147.00

			Skill Factor	

Flush door 35 mm thick
Red Oak veneered finish,
hardwood lipped, size

1981×610 mm	1 no	1:30	131.50	6	145.50
1981×686 mm	1 no	1:30	131.50	6	145.50
1981×762 mm	1 no	1:30	132.00	6	146.00
1981×838 mm	1 no	1:30	147.00	6	161.00

Internal flush premium softwood doors

Premium softwood
one-panel door, 35 mm thick,
(style I10) size

1981×686 mm	1 no	1:30	95.55	6	110.00
1981×762 mm	1 no	1:30	97.00	6	111.00
1981×838 mm	1 no	1:30	100.00	6	114.00
2032×813 mm	1 no	1:30	100.00	6	114.00

Premium softwood
ten-panel door, 35 mm thick,
(style ISC) size

1981×686 mm	1 no	1:30	180.50	6	195.00
1981×762 mm	1 no	1:30	182.00	6	196.00
1981×838 mm	1 no	1:30	189.00	6	203.00

Premium softwood two-panel
door, 35 mm thick,
(style I2XG) size

1981×762 mm	1 no	1:30	182.00	6	196.00
1981×838 mm	1 no	1:30	189.00	6	203.00
2032×813 mm	1 no	1:30	115.50	6	130.00
2040×726 mm	1 no	1:30	116.50	6	130.00
2040×826 mm	1 no	1:30	116.50	6	130.00

Premium softwood two-panel
door, 35 mm thick,
(style I2XGG) size

1981×762 mm	1 no	1:30	116.50	6	130.00
1981×838 mm	1 no	1:30	110.00	6	125.00
2032×813 mm	1 no	1:30	116.50	6	130.00
2040×726 mm	1 no	1:30	110.00	6	125.00
2040×826 mm	1 no	1:30	110.00	6	125.00

						Skill Factor	

	⫘	⏱	£	Skill Factor	♦

Premium softwood four-panel door, 35 mm thick, (style I4XGG) size

1981 × 686 mm	1 no	1:30	138.50	6	152.00
1981 × 762 mm	1 no	1:30	144.00	6	160.00
1981 × 838 mm	1 no	1:30	150.00	6	165.00

Premium softwood four-panel door, 35 mm thick, (style I4XPP) size

1981 × 686 mm	1 no	1:30	165.00	6	180.00
1981 × 762 mm	1 no	1:30	166.00	6	180.00
1981 × 838 mm	1 no	1:30	175.00	6	190.00

Premium softwood four-panel door, 35 mm thick, (style I50) size

1981 × 686 mm	1 no	1:30	113.00	6	130.00
1981 × 762 mm	1 no	1:30	113.00	6	130.00
1981 × 838 mm	1 no	1:30	117.00	6	132.00
2032 × 813 mm	1 no	1:30	117.00	6	132.00

Premium softwood four-panel door, 35 mm thick, (style ISA) size

1981 × 686 mm	1 no	1:30	170.00	6	185.00
1981 × 762 mm	1 no	1:30	172.00	6	185.00
1981 × 838 mm	1 no	1:30	183.00	6	190.00
2032 × 813 mm	1 no	1:30	183.00	6	190.00
2040 × 726 mm	1 no	1:30	180.00	6	185.00
2040 × 826 mm	1 no	1:30	185.00	6	200.00

Internal flush premium softwood doors with clear glass pack supplied

Premium softwood ten-panel door, 35 mm thick, (style ISC) with clear glass pack supplied, size

1981 × 686 mm	1 no	1:30	285.00	6	300.00
1981 × 762 mm	1 no	1:30	290.00	6	305.00
1981 × 838 mm	1 no	1:30	318.00	6	335.00

				Skill Factor	

Premium softwood four-panel door, 35 mm thick, (style ISA) with clear glass pack supplied, size					
1981×686 mm	1 no	1:30	270.00	6	285.00
1981×762 mm	1 no	1:30	283.00	6	300.00
1981×838 mm	1 no	1:30	300.00	6	315.00
2032×813 mm	1 no	1:30	300.00	6	315.00
2040×726 mm	1 no	1:30	295.00	6	310.00
2040×826 mm	1 no	1:30	306.00	6	320.00

Internal flush premium softwood doors with obscure glass pack supplied

Premium softwood ten-panel door, 35 mm thick, (style ISC) with obscure glass pack supplied, size					
1981×686 mm	1 no	1:30	306.00	6	320.00
1981×762 mm	1 no	1:30	314.00	6	330.00
1981×838 mm	1 no	1:30	340.00	6	360.00
Premium softwood four-panel door, 35 mm thick, (style ISA) with obscure glass pack supplied, size					
1981×686 mm	1 no	1:30	291.00	6	305.00
1981×762 mm	1 no	1:30	306.00	6	320.00
1981×838 mm	1 no	1:30	322.00	6	338.00
2032×813 mm	1 no	1:30	320.00	6	335.00
2040×726 mm	1 no	1:30	318.00	6	335.00
2040×826 mm	1 no	1:30	330.00	6	345.00
Flush door 44 mm thick hardboard faced both sides, half-hour fire door, size					
1981×686 mm	1 no	1:30	73.50	6	88.00
1981×762 mm	1 no	1:30	73.50	6	88.00
1981×838 mm	1 no	1:30	77.00	6	92.00
2032×813 mm	1 no	1:30	77.00	6	92.00
2040×526 mm	1 no	1:30	75.50	6	90.00
2040×626 mm	1 no	1:30	75.50	6	90.00
2040×726 mm	1 no	1:30	75.50	6	90.00
2040×826 mm	1 no	1:30	75.00	6	90.00

				Skill Factor	

Flush door 44 mm thick African Maple veneered finish, half-hour fire door, size					
1981×686 mm	1 no	1:30	147.00	6	150.00
1981×762 mm	1 no	1:30	147.00	6	150.00
1981×838 mm	1 no	1:30	152.50	6	166.00
2032×864 mm	1 no	1:30	173.50	6	188.00
Flush door 44 mm thick Sapele veneered finish, half-hour fire door, size					
1981×610 mm	1 no	1:30	114.00	6	128.00
1981×686 mm	1 no	1:30	114.00	6	128.00
1981×762 mm	1 no	1:30	114.00	6	128.00
1981×838 mm	1 no	1:30	118.00	6	132.00
2032×864 mm	1 no	1:30	118.00	6	132.00
Flush door 44 mm thick Beech veneered finish, half-hour fire door, size					
1981×686 mm	1 no	1:30	210.50	6	225.00
1981×762 mm	1 no	1:30	210.50	6	225.00
1981×838 mm	1 no	1:30	218.00	6	232.00
Flush door 44 mm thick White Oak veneered finish, half-hour fire door, size					
1981×686 mm	1 no	1:30	216.50	6	230.00
1981×762 mm	1 no	1:30	216.50	6	230.00
1981×838 mm	1 no	1:30	224.00	6	238.00
Flush door 44 mm thick Cherry veneered finish, half-hour fire door, size					
1981×686 mm	1 no	1:30	223.00	6	237.00
1981×762 mm	1 no	1:30	223.00	6	237.00
1981×838 mm	1 no	1:30	230.50	6	245.00
Flush door 44 mm thick Ash veneered finish, half-hour fire door, size					
1981×686 mm	1 no	1:30	196.00	6	210.00
1981×762 mm	1 no	1:30	196.00	6	210.00
1981×838 mm	1 no	1:30	204.00	6	218.00

			Skill Factor	

Flush door 44 mm thick
Red Oak veneered finish,
half-hour fire door, size

1981 × 686 mm	1 no	1:30	216.50	6	230.00
1981 × 762 mm	1 no	1:30	216.50	6	230.00
1981 × 838 mm	1 no	1:30	224.00	6	238.00

**Internal moulded
panel doors**

White primed moulded
panel six-panel door,
35 mm thick, size

1981 × 457 mm	1 no	1:30	78.50	6	92.00
1981 × 533 mm	1 no	1:30	78.50	6	92.00
1981 × 610 mm	1 no	1:30	78.50	6	92.00
1981 × 686 mm	1 no	1:30	79.00	6	92.00
1981 × 762 mm	1 no	1:30	79.00	6	92.00
1981 × 838 mm	1 no	1:30	79.00	6	95.00
2032 × 864 mm	1 no	1:30	93.00	6	98.00

White primed middle-weight
moulded panel six-panel
door, 35 mm thick, size

1981 × 457 mm	1 no	1:30	97.50	6	114.00
1981 × 533 mm	1 no	1:30	97.50	6	114.00
1981 × 610 mm	1 no	1:30	97.50	6	114.00
1981 × 686 mm	1 no	1:30	97.50	6	114.00
1981 × 762 mm	1 no	1:30	104.00	6	120.00
1981 × 838 mm	1 no	1:30	106.00	6	122.00
2032 × 864 mm	1 no	1:30	114.00	6	130.00

White primed heavy-weight
moulded panel six-panel
door, 40 mm thick, size

1981 × 457 mm	1 no	1:30	120.00	6	135.00
1981 × 533 mm	1 no	1:30	120.00	6	135.00
1981 × 610 mm	1 no	1:30	120.00	6	135.00
1981 × 686 mm	1 no	1:30	120.00	6	135.00
1981 × 762 mm	1 no	1:30	120.00	6	135.00
1981 × 838 mm	1 no	1:30	125.00	6	140.00
2032 × 864 mm	1 no	1:30	130.00	6	145.00

			Skill Factor	

White primed moulded panel two-panel door, 35 mm thick, wood grain finish, size					
1981×457 mm	1 no	1:30	82.50	6	98.00
1981×533 mm	1 no	1:30	82.50	6	98.00
1981×610 mm	1 no	1:30	82.50	6	98.00
1981×686 mm	1 no	1:30	82.50	6	98.00
1981×762 mm	1 no	1:30	82.50	6	98.00
1981×838 mm	1 no	1:30	92.00	6	105.00
White primed middle-weight moulded panel two-panel door, 35 mm thick, wood grain finish, size					
1981×457 mm	1 no	1:30	100.00	6	115.00
1981×533 mm	1 no	1:30	100.00	6	115.00
1981×610 mm	1 no	1:30	100.00	6	115.00
1981×686 mm	1 no	1:30	100.00	6	115.00
1981×762 mm	1 no	1:30	100.00	6	115.00
1981×838 mm	1 no	1:30	108.00	6	125.00
2032×864 mm	1 no	1:30	118.00	6	132.00
White primed moulded panel four-panel door, 35 mm thick, wood grain finish, size					
1981×457 mm	1 no	1:30	77.50	6	92.00
1981×533 mm	1 no	1:30	77.50	6	92.00
1981×610 mm	1 no	1:30	77.50	6	92.00
1981×686 mm	1 no	1:30	77.50	6	92.00
1981×762 mm	1 no	1:30	80.00	6	95.00
1981×838 mm	1 no	1:30	87.00	6	102.00
2032×864 mm	1 no	1:30	87.00	6	102.00
White primed middle-weight moulded panel four-panel door, 35 mm thick, wood grain finish, size					
1981×457 mm	1 no	1:30	98.50	6	112.00
1981×533 mm	1 no	1:30	98.50	6	112.00
1981×610 mm	1 no	1:30	98.50	6	112.00
1981×686 mm	1 no	1:30	98.50	6	112.00
1981×762 mm	1 no	1:30	103.00	6	118.00
1981×838 mm	1 no	1:30	105.00	6	120.00

	📏	🕐	£	Skill Factor	🔨
White primed moulded panel six-panel door, 35 mm thick, wood grain finish, size					
1981×457 mm	1 no	1:30	77.50	6	92.00
1981×533 mm	1 no	1:30	77.50	6	92.00
1981×610 mm	1 no	1:30	77.50	6	92.00
1981×686 mm	1 no	1:30	77.50	6	92.00
1981×762 mm	1 no	1:30	80.00	6	95.00
1981×838 mm	1 no	1:30	87.00	6	102.00
2032×864 mm	1 no	1:30	87.00	6	102.00
White primed middle-weight moulded panel six-panel door, 35 mm thick, wood grain finish, size					
1981×457 mm	1 no	1:30	98.50	6	112.00
1981×533 mm	1 no	1:30	98.50	6	112.00
1981×610 mm	1 no	1:30	98.50	6	112.00
1981×686 mm	1 no	1:30	98.50	6	112.00
1981×762 mm	1 no	1:30	103.00	6	118.00
1981×838 mm	1 no	1:30	105.00	6	120.00
2032×864 mm	1 no	1:30	113.00	6	130.00

Internal pine doors

Clear pine, six-panel door, 35 mm thick, size					
1981×610 mm	1 no	1:30	112.00	6	125.00
1981×686 mm	1 no	1:30	115.00	6	130.00
1981×762 mm	1 no	1:30	116.00	6	130.00
1981×838 mm	1 no	1:30	121.00	6	135.00
Clear pine, four-panel door, 35 mm thick, size					
1981×610 mm	1 no	1:30	112.00	6	125.00
1981×686 mm	1 no	1:30	115.00	6	130.00
1981×762 mm	1 no	1:30	116.00	6	130.00
1981×838 mm	1 no	1:30	121.00	6	135.00
Knotty pine, six-panel door, 35 mm thick, size					
1981×610 mm	1 no	1:30	101.00	6	115.00
1981×686 mm	1 no	1:30	103.00	6	118.00
1981×762 mm	1 no	1:30	105.00	6	120.00
1981×838 mm	1 no	1:30	109.00	6	124.00

	📏	🕐	💷	Skill Factor	🔶
Knotty pine, four-panel door, 35 mm thick, size					
1981 × 610 mm	1 no	1:30	101.00	6	115.00
1981 × 686 mm	1 no	1:30	103.00	6	118.00
1981 × 762 mm	1 no	1:30	105.00	6	120.00
1981 × 838 mm	1 no	1:30	109.00	6	124.00
Clear pine, fifteen-panel glazed door, 35 mm thick, size					
1981 × 610 mm	1 no	1:30	250.00	6	275.00
1981 × 686 mm	1 no	1:30	280.00	6	300.00
1981 × 762 mm	1 no	1:30	282.00	6	305.00
1981 × 838 mm	1 no	1:30	288.00	6	310.00
Internal framed, ledged and braced doors					
44 mm thick framed, ledged and braced door					
1981 × 686 mm	1 no	1:30	245.00	6	265.00
1981 × 762 mm	1 no	1:30	245.00	6	265.00
1981 × 838 mm	1 no	1:30	260.00	6	280.00
Internal louvred wardrobe doors					
27 mm thick louvred wardrobe doors 305 mm wide, height					
457 mm	1 no	1:00	11.15	6	20.00
610 mm	1 no	1:00	14.80	6	25.00
762 mm	1 no	1:00	17.40	6	28.00
914 mm	1 no	1:00	20.00	6	30.00
1067 mm	1 no	1:00	22.95	6	33.00
1219 mm	1 no	1:00	25.80	6	35.00
1372 mm	1 no	1:00	30.00	6	40.00
1524 mm	1 no	1:15	33.60	6	48.00
1676 mm	1 no	1:15	36.40	6	50.00
1829 mm	1 no	1:15	38.90	6	52.00
1981 mm	1 no	1:15	41.80	6	55.00
27 mm thick louvred wardrobe doors 381 mm wide, height					
457 mm	1 no	1:00	12.90	6	22.00
610 mm	1 no	1:00	16.25	6	28.00
762 mm	1 no	1:00	19.50	6	30.00
914 mm	1 no	1:00	23.00	6	33.00

	📏	🕐	£	Skill Factor	⚡
1067 mm	1 no	1:00	26.25	6	35.00
1219 mm	1 no	1:00	29.50	6	40.00
1372 mm	1 no	1:00	34.50	6	45.00
1524 mm	1 no	1:15	38.50	6	50.00
1676 mm	1 no	1:15	41.90	6	52.00
1829 mm	1 no	1:15	44.75	6	60.00
1981 mm	1 no	1:15	48.00	6	62.00
27 mm thick louvred wardrobe doors 457 mm wide, height					
457 mm	1 no	1:00	14.75	6	25.00
610 mm	1 no	1:00	18.50	6	30.00
762 mm	1 no	1:00	22.75	6	32.00
914 mm	1 no	1:00	26.00	6	35.00
1067 mm	1 no	1:00	29.50	6	40.00
1219 mm	1 no	1:00	33.50	6	45.00
1372 mm	1 no	1:00	39.25	6	50.00
1524 mm	1 no	1:15	43.50	6	54.00
1676 mm	1 no	1:15	47.50	6	58.00
1829 mm	1 no	1:15	50.50	6	60.00
1981 mm	1 no	1:15	54.25	6	65.00
27 mm thick louvred wardrobe doors 457 mm wide, height					
457 mm	1 no	1:00	16.50	6	25.00
610 mm	1 no	1:00	20.75	6	30.00
762 mm	1 no	1:00	24.75	6	35.00
914 mm	1 no	1:00	29.00	6	40.00
1067 mm	1 no	1:00	33.00	6	43.00
1219 mm	1 no	1:00	37.00	6	48.00
1372 mm	1 no	1:00	44.00	6	54.00
1524 mm	1 no	1:15	48.50	6	64.00
1676 mm	1 no	1:15	53.00	6	68.00
1829 mm	1 no	1:15	56.00	6	70.00
1981 mm	1 no	1:15	60.00	6	75.00
27 mm thick louvred wardrobe doors 457 mm wide, height					
457 mm	1 no	1:00	18.50	6	30.00
610 mm	1 no	1:00	23.00	6	33.00
762 mm	1 no	1:00	27.50	6	38.00
914 mm	1 no	1:00	32.00	6	42.00
1067 mm	1 no	1:00	36.50	6	52.00
1219 mm	1 no	1:00	41.00	6	45.00

				Skill Factor	
1372 mm	1 no	1:00	48.50	6	60.00
1524 mm	1 no	1:15	53.50	6	68.00
1676 mm	1 no	1:15	58.50	6	70.00
1829 mm	1 no	1:15	62.00	6	77.00
1981 mm	1 no	1:15	66.50	6	80.00

External hardwood panel doors

Hardwood six-panel door, 44 mm thick, factory clear glazed, (style HKXT), size

1981 × 762 mm	1 no	1:45	545.00	6	575.00
1981 × 838 mm	1 no	1:45	550.00	6	580.00
2032 × 813 mm	1 no	1:45	555.00	6	585.00

Hardwood six-panel door, 44 mm thick, factory obscure glazed, (style HKXT), size

1981 × 762 mm	1 no	1:45	565.00	6	590.00
1981 × 838 mm	1 no	1:45	572.00	6	600.00
2032 × 813 mm	1 no	1:45	572.00	6	600.00

Hardwood fifteen-panel door, 44 mm thick, factory clear glazed, (style HESA), size

1981 × 762 mm	1 no	1:45	635.00	6	665.00
1981 × 838 mm	1 no	1:45	645.00	6	675.00
2032 × 813 mm	1 no	1:45	645.00	6	675.00

Hardwood fifteen-panel door, 44 mm thick, factory obscure glazed, (style HESA), size

1981 × 762 mm	1 no	1:45	696.00	6	725.00
1981 × 838 mm	1 no	1:45	705.00	6	735.00
2032 × 813 mm	1 no	1:45	705.00	6	735.00

Hardwood two-panel door, 44 mm thick, factory clear glazed, (style HX2GG), size

1981 × 762 mm	1 no	1:45	344.00	6	375.00
1981 × 838 mm	1 no	1:45	350.00	6	380.00
2032 × 813 mm	1 no	1:45	350.00	6	380.00

				Skill Factor	

Hardwood two-panel door,
44 mm thick, factory
obscure glazed,
(style HX2GG), size

1981×762 mm	1 no	1:45	363.00	6	395.00
1981×838 mm	1 no	1:45	368.00	6	400.00
2032×813 mm	1 no	1:45	368.00	6	400.00

Hardwood two-panel door,
44 mm thick, factory
clear glazed, (style HX2GGS),
size

1981×762 mm	1 no	1:45	392.00	6	425.00
1981×838 mm	1 no	1:45	398.00	6	430.00
2032×813 mm	1 no	1:45	398.00	6	430.00

Hardwood two-panel door,
44 mm thick, factory
obscure glazed,
(style HX2GGS), size

1981×762 mm	1 no	1:45	411.00	6	440.00
1981×838 mm	1 no	1:45	418.00	6	450.00
2032×813 mm	1 no	1:45	418.00	6	450.00

**External softwood
panel doors**

Softwood two-panel door,
44 mm thick, factory clear
glazed, (style E2XGG), size

1981×762 mm	1 no	1:45	269.00	6	300.00
1981×838 mm	1 no	1:45	289.00	6	320.00
2032×813 mm	1 no	1:45	289.00	6	320.00
2057×838 mm	1 no	1:45	311.00	6	340.00
2083×864 mm	1 no	1:45	313.00	6	345.00

Softwood two-panel door,
44 mm thick, factory
obscure glazed,
(style E2XGG), size

1981×762 mm	1 no	1:45	291.00	6	320.00
1981×838 mm	1 no	1:45	314.00	6	345.00
2032×813 mm	1 no	1:45	314.00	6	345.00
2057×838 mm	1 no	1:45	339.00	6	370.00
2083×864 mm	1 no	1:45	340.00	6	370.00

				Skill Factor	

Softwood fifteen-panel door, 44 mm thick, factory clear glazed, (style ESA), size					
1981 × 762 mm	1 no	1:45	435.00	6	475.00
1981 × 838 mm	1 no	1:45	457.00	6	490.00
2032 × 813 mm	1 no	1:45	457.00	6	490.00
Softwood fifteen-panel door, 44 mm thick, factory obscure glazed, (style ESA), size					
1981 × 762 mm	1 no	1:45	460.00	6	490.00
1981 × 838 mm	1 no	1:45	485.00	6	515.00
2032 × 813 mm	1 no	1:45	485.00	6	515.00
Softwood single-panel door, 44 mm thick, factory clear glazed, (style PATT 10), size					
1981 × 762 mm	1 no	1:45	277.00	6	310.00
1981 × 838 mm	1 no	1:45	298.00	6	330.00
2032 × 813 mm	1 no	1:45	298.00	6	330.00
Softwood single-panel door, 44 mm thick, factory obscure glazed, (style PATT 10), size					
1981 × 762 mm	1 no	1:45	301.00	6	330.00
1981 × 838 mm	1 no	1:45	326.00	6	355.00
2032 × 813 mm	1 no	1:45	326.00	6	355.00
Softwood three-panel door, 44 mm thick, factory clear glazed, (style PATT 80), size					
1981 × 762 mm	1 no	1:45	303.00	6	335.00
1981 × 838 mm	1 no	1:45	316.00	6	345.00
2032 × 813 mm	1 no	1:45	321.00	6	350.00
Softwood three-panel door, 44 mm thick, factory obscure glazed, (style PATT 80), size					
1981 × 762 mm	1 no	1:45	319.00	6	350.00
1981 × 838 mm	1 no	1:45	334.00	6	365.00
2032 × 813 mm	1 no	1:45	340.00	6	370.00

These prices may need adjusting to suit the price levels in your area.
See page xi in the introduction on how to adapt them
for your particular part of the country

	▤	🕐	£	Skill Factor	♦
Softwood four-panel door, 44 mm thick, factory clear glazed, (style PATT 50), size					
1981×762 mm	1 no	1:45	338.00	6	370.00
1981×838 mm	1 no	1:45	353.00	6	385.00
2032×813 mm	1 no	1:45	360.00	6	390.00
Softwood four-panel door, 44 mm thick, factory obscure glazed, (style PATT 50), size					
1981×762 mm	1 no	1:45	361.00	6	390.00
1981×838 mm	1 no	1:45	377.00	6	410.00
2032×813 mm	1 no	1:45	386.00	6	415.00
Softwood four-panel door, 44 mm thick, factory clear glazed, (style PATT E4XG), size					
1981×762 mm	1 no	1:45	294.00	6	325.00
1981×838 mm	1 no	1:45	310.00	6	340.00
Softwood four-panel door, 44 mm thick, factory obscure glazed, (style PATT E4XG), size					
1981×762 mm	1 no	1:45	306.00	6	335.00
1981×838 mm	1 no	1:45	324.00	6	355.00
Softwood single-panel door, 44 mm thick, factory clear glazed, (style PATT E2XG), size					
1981×762 mm	1 no	1:45	203.00	6	235.00
1981×838 mm	1 no	1:45	215.00	6	245.00
2032×813 mm	1 no	1:45	215.00	6	145.00
Softwood single-panel door, 44 mm thick, factory obscure glazed, (style PATT E2XG), size					
1981×762 mm	1 no	1:45	210.00	6	240.00
1981×838 mm	1 no	1:45	222.00	6	255.00
2032×813 mm	1 no	1:45	222.00	6	255.00

These prices may need adjusting to suit the price levels in your area.
See page xi in the introduction on how to adapt them
for your particular part of the country

DOORS

External redwood panel doors					
Redwood six-panel door, 44 mm thick, factory clear glazed, (style REKXT), size					
1981×762 mm	1 no	1:45	330.00	6	365.00
1981×838 mm	1 no	1:45	332.00	6	370.00
2032×813 mm	1 no	1:45	332.00	6	370.00
Softwood three-panel door, 44 mm thick, factory obscure glazed, (style REKXT), size					
1981×762 mm	1 no	1:45	350.00	6	380.00
1981×838 mm	1 no	1:45	352.00	6	385.00
2032×813 mm	1 no	1:45	352.00	6	385.00
Redwood single-panel door, 44 mm thick, factory clear glazed, (style REXG), size					
1981×762 mm	1 no	1:45	178.00	6	210.00
1981×838 mm	1 no	1:45	180.00	6	215.00
2032×813 mm	1 no	1:45	180.00	6	215.00
Softwood single-panel door, 44 mm thick, factory obscure glazed, (style REXG), size					
1981×762 mm	1 no	1:45	190.00	6	220.00
1981×838 mm	1 no	1:45	192.00	6	225.00
2032×813 mm	1 no	1:45	192.00	6	225.00
Redwood single-panel door, 44 mm thick, factory clear glazed, (style RE10), size					
1981×762 mm	1 no	1:45	240.00	6	270.00
1981×838 mm	1 no	1:45	257.00	6	290.00
2032×813 mm	1 no	1:45	257.00	6	290.00
Softwood single-panel door, 44 mm thick, factory obscure glazed, (style RE10), size					
1981×762 mm	1 no	1:45	258.00	6	290.00
1981×838 mm	1 no	1:45	277.00	6	310.00
2032×813 mm	1 no	1:45	277.00	6	310.00

				Skill Factor	

Redwood two-panel door, 44 mm thick, factory clear glazed, (style RE2XGG), size					
1981 × 762 mm	1 no	1:45	231.00	6	260.00
1981 × 838 mm	1 no	1:45	232.00	6	260.00
2032 × 813 mm	1 no	1:45	232.00	6	260.00
2057 × 838 mm	1 no	1:45	252.00	6	285.00
Softwood single-panel door, 44 mm thick, factory obscure glazed, (style RE2XGG), size					
1981 × 762 mm	1 no	1:45	249.00	6	280.00
1981 × 838 mm	1 no	1:45	251.00	6	280.00
2032 × 813 mm	1 no	1:45	251.00	6	280.00
2057 × 838 mm	1 no	1:45	272.00	6	305.00
External Redwood boarded doors					
Redwood framed and braced doors, 44 mm thick, size					
1981 × 610 mm	1 no	1:45	77.50	6	92.00
1981 × 686 mm	1 no	1:45	77.50	6	92.00
1981 × 762 mm	1 no	1:45	80.00	6	95.00
1981 × 838 mm	1 no	1:45	87.00	6	102.00
2032 × 813 mm	1 no	1:45	87.00	6	102.00
Redwood, ledged framed and braced doors, 44 mm thick, size					
1981 × 686 mm	1 no	1:45	77.50	6	92.00
1981 × 762 mm	1 no	1:45	80.00	6	95.00
1981 × 838 mm	1 no	1:45	87.00	6	102.00
2032 × 813 mm	1 no	1:45	87.00	6	102.00
Redwood stable door in two leaves, 44 mm thick, overall size					
1981 × 762 mm	1 no	1:45	80.00	6	95.00
1981 × 838 mm	1 no	1:45	87.00	6	102.00
2032 × 813 mm	1 no	1:45	87.00	6	102.00
Garage doors					
Pair of Redwood garage doors, overall size					
2134 × 1981 mm	1 pr	8:00	386.00	6	600.00
2134 × 2134 mm	1 pr	8:00	398.00	6	625.00

			£	Skill Factor	
Hardwood single-panel up-and-over door, manually operated, overall size					
2134×1981 mm	1 no	10:00	1050.00	6	1500.00
2286×2134 mm	1 no	10:00	1200.00	6	1750.00
4627×2134 mm	1 no	16:00	2500.00	6	3000.00
Hardwood single-panel up-and-over door, remote controlled operated, overall size					
2134×1981 mm	1 no	16:00	1600.00	6	2200.00
2286×2134 mm	1 no	16:00	1800.00	6	2400.00
4627×2134 mm	1 no	24:00	3000.00	6	3750.00
Steel roller shutter door, manually operated, overall size					
2134×1981 mm	1 no	16:00	700.00	6	1300.00
2438×2134 mm	1 no	16:00	800.00	6	1500.00
4627×2134 mm	1 no	24:00	1350.00	6	2000.00
Steel roller shutter door, remote control operated, overall size					
2134×1981 mm	1 no	16:00	1500.00	6	2100.00
2438×2134 mm	1 no	16:00	1650.00	6	2250.00
4627×2134 mm	1 no	24:00	2100.00	6	2750.00

WINDOWS

Although PVCU windows have made inroads into the domestic market in recent years, there is still a strong demand for both traditional softwood and hardwood windows. You can buy windows directly from the suppliers, already single or double glazed and in a wide variety of sizes and designs.

Most of the information on windows in this section is based on the information supplied by Jeld-Wen Uk Ltd, Doncaster (01302 394000) and the reference numbers are those that appear in their catalogue.

These prices may need adjusting to suit the price levels in your area.
See page xi in the introduction on how to adapt them
for your particular part of the country

	📏	🕐	£	Skill Factor	🔻£
Softwood windows					
Plain casement windows, overall size					
630 × 750 mm	1 no	1:30	118.00	6	148.00
630 × 900 mm	1 no	1:30	126.00	6	156.00
630 × 1050 mm	1 no	1:30	128.00	6	158.00
630 × 1200 mm	1 no	1:45	130.00	6	170.00
630 × 1350 mm	1 no	1:45	145.00	6	185.00
1200 × 750 mm	1 no	1:30	152.00	6	192.00
1200 × 900 mm	1 no	1:30	156.00	6	196.00
1200 × 1050 mm	1 no	1:30	160.00	6	200.00
1200 × 1200 mm	1 no	1:45	165.00	6	215.00
1200 × 1350 mm	1 no	1:45	182.00	6	232.00
1770 × 750 mm	1 no	1:45	225.00	6	275.00
1770 × 900 mm	1 no	1:45	230.00	6	280.00
1770 × 1050 mm	1 no	1:45	232.00	6	282.00
1770 × 1200 mm	1 no	2:00	240.00	6	300.00
1770 × 1350 mm	1 no	2:00	270.00	6	330.00
2339 × 900 mm	1 no	2:00	290.00	6	350.00
2339 × 1050 mm	1 no	2:00	300.00	6	360.00
2339 × 1200 mm	1 no	2:00	312.00	6	372.00
2339 × 1350 mm	1 no	2:00	332.00	6	392.00
Hardwood windows					
Plain casement windows, overall size					
630 × 750 mm	1 no	1:30	220.00	6	250.00
630 × 900 mm	1 no	1:30	228.00	6	258.00
630 × 1050 mm	1 no	1:30	238.00	6	268.00
630 × 1200 mm	1 no	1:45	248.00	6	288.00
630 × 1350 mm	1 no	1:45	266.00	6	306.00
1200 × 750 mm	1 no	1:30	282.00	6	322.00
1200 × 900 mm	1 no	1:30	296.00	6	336.00
1200 × 1050 mm	1 no	1:30	306.00	6	346.00
1200 × 1200 mm	1 no	1:45	320.00	6	370.00
1200 × 1350 mm	1 no	1:45	341.00	6	391.00
1770 × 750 mm	1 no	1:45	428.00	6	478.00
1770 × 900 mm	1 no	1:45	448.00	6	498.00
1770 × 1050 mm	1 no	1:45	466.00	6	516.00

			£	Skill Factor	£
1770×1200 mm	1 no	2:00	485.00	6	545.00
1770×1350 mm	1 no	2:00	520.00	6	580.00
2339×900 mm	1 no	2:00	566.00	6	626.00
2339×1050 mm	1 no	2:00	580.00	6	640.00
2339×1200 mm	1 no	2:00	606.00	6	666.00
2339×1350 mm	1 no	2:00	646.00	6	706.00

Softwood windows, factory glazed

Plain casement windows, overall size

630×750 mm	1 no	1:30	170.00	6	210.00
630×900 mm	1 no	1:30	183.00	6	223.00
630×1050 mm	1 no	1:30	190.00	6	230.00
630×1200 mm	1 no	1:45	200.00	6	250.00
630×1350 mm	1 no	1:45	228.00	6	278.00
1200×750 mm	1 no	1:30	266.00	6	316.00
1200×900 mm	1 no	1:30	283.00	6	333.00
1200×1050 mm	1 no	1:30	298.00	6	348.00
1200×1200 mm	1 no	1:45	321.00	6	381.00
1200×1350 mm	1 no	1:45	360.00	6	420.00
1770×750 mm	1 no	1:45	390.00	6	450.00
1770×900 mm	1 no	1:45	415.00	6	475.00
1770×1050 mm	1 no	1:45	435.00	6	495.00
1770×1200 mm	1 no	2:00	469.00	6	539.00
1770×1350 mm	1 no	2:00	530.00	6	600.00
2339×900 mm	1 no	2:00	540.00	6	610.00
2339×1050 mm	1 no	2:00	579.00	6	639.00
2339×1200 mm	1 no	2:00	630.00	6	690.00
2339×1350 mm	1 no	2:00	687.00	6	747.00

Hardwood windows, factory glazed

Plain casement windows, overall size

630×750 mm	1 no	1:30	273.00	6	313.00
630×900 mm	1 no	1:30	288.00	6	328.00
630×1050 mm	1 no	1:30	303.00	6	343.00
630×1200 mm	1 no	1:45	320.00	6	470.00
630×1350 mm	1 no	1:45	350.00	6	400.00

	≣	🕐	£	Skill Factor	♠
1200×750 mm	1 no	1:30	396.00	6	346.00
1200×900 mm	1 no	1:30	423.00	6	473.00
1200×1050 mm	1 no	1:30	450.00	6	500.00
1200×1200 mm	1 no	1:45	478.00	6	538.00
1200×1350 mm	1 no	1:45	520.00	6	580.00
1770×750 mm	1 no	1:45	593.00	6	653.00
1770×900 mm	1 no	1:45	635.00	6	695.00
1770×1050 mm	1 no	2:00	670.00	6	730.00
1770×1200 mm	1 no	2:00	715.00	6	785.00
1770×1350 mm	1 no	2:00	780.00	6	850.00
2339×900 mm	1 no	2:00	820.00	6	890.00
2339×1050 mm	1 no	2:00	860.00	6	930.00
2339×1200 mm	1 no	2:00	922.00	6	992.00
2339×1350 mm	1 no	2:00	1000.00	6	1070.00

Softwood sliding sash windows

Sliding sash windows fell out of favour about 50 years ago but are now becoming popular again. The following costs are for windows with and without glazing bars and include for glazing at the factory.

	≣	🕐	£	Skill Factor	♠
Softwood windows, without glazing bars, factory glazed					
Sliding sash windows, overall size					
410×1050 mm	1 no	1:30	495.00	6	545.00
410×1350 mm	1 no	1:30	510.00	6	560.00
410×1650 mm	1 no	1:45	536.00	6	586.00
635×1050 mm	1 no	1:30	508.00	6	558.00
635×1350 mm	1 no	1:45	528.00	6	578.00
635×1650 mm	1 no	1:45	560.00	6	610.00
860×1050 mm	1 no	1:45	535.00	6	695.00
860×1350 mm	1 no	1:45	562.00	6	662.00
860×1650 mm	1 no	1:45	618.00	6	678.00

	⊞	🕐	£	Skill Factor	🔨£
1085×1050 mm	1 no	1:45	570.00	6	630.00
1085×1350 mm	1 no	1:45	617.00	6	617.00
1085×1650 mm	1 no	1:45	670.00	6	730.00
1670×1050 mm	1 no	1:45	1005.00	6	1075.00
1670×1350 mm	1 no	1:45	1045.00	6	1115.00
1670×1650 mm	1 no	2:00	1126.00	6	1196.00
1715×1050 mm	1 no	1:45	928.00	6	1628.00
1715×1350 mm	1 no	1:45	987.00	6	1057.00
1715×1650 mm	1 no	2:00	1097.00	6	1167.00
1895×1050 mm	1 no	2:00	1043.00	6	1113.00
1895×1350 mm	1 no	2:00	1100.00	6	1170.00
1895×1650 mm	1 no	2:00	1177.00	6	1247.00

Softwood windows, with glazing bars, factory glazed

Sliding sash windows, overall size

410×1050 mm	1 no	1:30	522.00	6	572.00
410×1350 mm	1 no	1:30	560.00	6	610.00
410×1650 mm	1 no	1:45	610.00	6	660.00
635×1050 mm	1 no	1:30	598.00	6	648.00
635×1350 mm	1 no	1:45	662.00	6	712.00
635×1650 mm	1 no	1:45	748.00	6	798.00
860×1050 mm	1 no	1:45	685.00	6	745.00
860×1350 mm	1 no	1:45	765.00	6	825.00
860×1650 mm	1 no	1:45	887.00	6	947.00
1085×1050 mm	1 no	1:45	784.00	6	847.00
1085×1350 mm	1 no	1:45	886.00	6	946.00
1085×1650 mm	1 no	1:45	1094.00	6	1154.00
1670×1050 mm	1 no	1:45	1208.00	6	1278.00
1670×1350 mm	1 no	1:45	1345.00	6	1415.00
1670×1650 mm	1 no	2:00	1550.00	6	1620.00
1715×1050 mm	1 no	1:45	1140.00	6	1210.00
1715×1350 mm	1 no	1:45	1298.00	6	1368.00
1715×1650 mm	1 no	2:00	1540.00	6	1610.00

Rooflights

The higher price of property has generated an increase in the number of loft conversions. Gaining natural light to new rooms in lofts is achieved either by the provision of dormer windows or by roof windows. Here are some typical costs for the fixing of the roof windows excluding forming the opening in the roof and the peripheral costs. See *Loft conversions* on pages 125 to 126 for other costs.

					Skill Factor	
Rooflight overall size						
540 × 700 mm	1 no	2:00	370.00	6		410.00
540 × 980 mm	1 no	2:00	400.00	6		440.00
780 × 980 mm	1 no	2:30	440.00	6		490.00
780 × 1180 mm	1 no	2:45	470.00	6		530.00
1140 × 1180 mm	1 no	3:00	550.00	6		610.00
1340 × 980 mm	1 no	3:00	565.00	6		625.00

KITCHEN FITTINGS

There are two main types of kitchen fittings – flat-pack and ready-assembled. The flat-pack products are cheaper and their quality has increased enormously in the last few years. Any self-respecting DIYer should be able to put together flat-pack kitchen units providing they follow the instructions carefully and observe some common-sense rules. These include only opening one pack at a time and assembling the unit in a large cleared space.

Ready-assembled units can range from standard to luxury with matching prices! Imported Italian kitchens can be very expensive indeed, but the figures quoted in the tables below represent a middle of the range product available in your local DIY supermarket.

The times quoted for the flat-pack units is for the assembly and the fitting, but only the fitting for the ready-assembled units. The dimensions are in the order of width × depth × height.

These prices may need adjusting to suit the price levels in your area.
See page xi in the introduction on how to adapt them
for your particular part of the country

				Skill Factor	

Flat-pack units

Floor cupboards, size					
300 × 600 × 870 mm	1 no	2:00	82.00	6	142.00
400 × 600 × 870 mm	1 no	2:00	85.00	6	145.00
500 × 600 × 870 mm	1 no	2:00	92.00	6	152.00
600 × 600 × 870 mm	1 no	2:00	98.00	6	158.00
800 × 600 × 870 mm	1 no	2:00	138.00	6	178.00
1000 × 600 × 870 mm	1 no	2:00	150.00	6	210.00
Floor cupboards with drawer, size					
300 × 600 × 870 mm	1 no	2:00	112.00	6	172.00
400 × 600 × 870 mm	1 no	2:00	122.00	6	182.00
500 × 600 × 870 mm	1 no	2:00	130.00	6	190.00
600 × 600 × 870 mm	1 no	2:00	140.00	6	200.00
800 × 600 × 870 mm	1 no	2:00	210.00	6	270.00
1000 × 600 × 870 mm	1 no	2:00	225.00	6	295.00
Wall cupboards, size					
300 × 300 × 720 mm	1 no	2:00	65.00	6	125.00
400 × 300 × 720 mm	1 no	2:00	70.00	6	130.00
500 × 300 × 720 mm	1 no	2:00	75.00	6	135.00
600 × 300 × 720 mm	1 no	2:00	80.00	6	140.00
800 × 300 × 720 mm	1 no	2:00	115.00	6	175.00
1000 × 300 × 720 mm	1 no	2:00	135.00	6	195.00
Sink unit, size					
1000 × 600 × 870 mm	1 no	2:00	205.00	6	265.00
Corner wall cupboards, size					
300 × 300 × 720 mm	1 no	2:00	100.00	6	120.00
Over hob unit size,					
600 × 300 × 360 mm	1 no	2:00	60.00	6	80.00
Oven housing unit size,					
600 × 600 × 2125 mm	1 no	2:00	260.00	6	300.00

Ready-assembled units

Floor cupboards, size					
300 × 600 × 870 mm	1 no	2:00	87.00	6	147.00
400 × 600 × 870 mm	1 no	2:00	90.00	6	150.00
500 × 600 × 870 mm	1 no	2:00	97.00	6	157.00
600 × 600 × 870 mm	1 no	2:00	103.00	6	163.00
800 × 600 × 870 mm	1 no	2:00	148.00	6	188.00
1000 × 600 × 870 mm	1 no	2:00	160.00	6	220.00

		⊞	'L'	£	Skill Factor	⚡
Floor cupboards with drawer, size						
300×600×870 mm		1 no	2:00	117.00	6	177.00
400×600×870 mm		1 no	2:00	127.00	6	187.00
500×600×870 mm		1 no	2:00	135.00	6	195.00
600×600×870 mm		1 no	2:00	145.00	6	205.00
800×600×870 mm		1 no	2:00	220.00	6	280.00
1000×600×870 mm		1 no	2:00	235.00	6	305.00
Wall cupboards, size						
300×300×720 mm		1 no	2:00	70.00	6	130.00
400×300×720 mm		1 no	2:00	75.00	6	135.00
500×300×720 mm		1 no	2:00	80.00	6	140.00
600×300×720 mm		1 no	2:00	85.00	6	145.00
800×300×720 mm		1 no	2:00	125.00	6	185.00
1000×300×720 mm		1 no	2:00	145.00	6	205.00
Sink unit, size						
1000×600×870 mm		1 no	2:00	215.00	6	275.00
Corner wall cupboards, size						
300×300×720 mm		1 no	2:00	110.00	6	130.00
Over hob unit size,						
600×300×360 mm		1 no	2:00	65.00	6	90.00
Oven housing unit size,						
600×600×2125 mm		1 no	2:00	270.00	6	310.00

The practice of changing the appearance of a kitchen by replacing unit doors and drawer fronts is becoming more popular because of the cost savings and the minimum disruption involved, compared to having the kitchen renewed.

The following costs are based upon the information supplied by Woodfit Ltd, Kem Mill, Whittle-le-Woods, Chorley, Lancashire, PR6 7EA (01257 266421).

		⊞	'L'	£	Skill Factor	⚡
Solid oak drawer fronts, size						
110×597 mm		1 no	0:45	22.00	6	37.00
140×297 mm		1 no	0:45	14.00	6	29.00
140×397 mm		1 no	0:45	18.00	6	33.00

	▥	🕐	£	Skill Factor	◈
140×497 mm	1 no	0:45	20.00	6	35.00
175×497 mm	1 no	0:45	24.00	6	39.00
175×597 mm	1 no	0:45	26.00	6	41.00
Solid oak doors, size					
450×597 mm	1 no	1:00	80.00	6	100.00
570×297 mm	1 no	1:00	57.00	6	57.00
570×447 mm	1 no	1:00	69.00	6	89.00
570×497 mm	1 no	1:00	69.00	6	89.00
715×297 mm	1 no	1:00	66.00	6	86.00
715×397 mm	1 no	1:00	71.00	6	91.00
Solid maple drawer fronts, size					
110×595 mm	1 no	0:45	20.00	6	35.00
140×295 mm	1 no	0:45	13.00	6	38.00
140×395 mm	1 no	0:45	15.00	6	40.00
140×495 mm	1 no	0:45	18.00	6	43.00
175×495 mm	1 no	0:45	21.00	6	46.00
175×595 mm	1 no	0:45	25.00	6	50.00
Solid maple doors, size					
450×595 mm	1 no	1:00	49.00	6	69.00
570×295 mm	1 no	1:00	38.00	6	58.00
570×395 mm	1 no	1:00	44.00	6	64.00
570×495 mm	1 no	1:00	44.00	6	64.00
570×595 mm	1 no	1:00	52.00	6	72.00
715×295 mm	1 no	1:00	42.00	6	82.00
Vintage pine drawer fronts, size					
110×597 mm	1 no	0:45	36.00	6	51.00
140×297 mm	1 no	0:45	23.00	6	38.00
140×397 mm	1 no	0:45	28.00	6	43.00
140×497 mm	1 no	0:45	32.00	6	47.00
175×497 mm	1 no	0:45	38.00	6	53.00
175×597 mm	1 no	0:45	42.00	6	57.00
Vintage pine doors, size					
450×597 mm	1 no	1:00	80.00	6	100.00
570×297 mm	1 no	1:00	58.00	6	78.00
570×447 mm	1 no	1:00	69.00	6	89.00
570×497 mm	1 no	1:00	69.00	6	89.00
715×297 mm	1 no	1:00	65.00	6	85.00
715×397 mm	1 no	1:00	70.00	6	90.00

WALL, FLOOR AND CEILING FINISHINGS

There are many DIY jobs that are hidden from view when completed, so that the appearance is not too important but the nature of work in this section means that the finished product will be on view permanently so it needs extra care and attention in carrying it out.

If you have not tried plastering before, you may be surprised to find that it is one of the hardest skills to master. If you intend having a go, start on a wall that is not a public view, behind a door for example, but not over the fireplace in the lounge that will be facing you every day for many years!

Laying floor and wall tiles can also be difficult but most of the DIYers can achieve a reasonable level of finish with a little practice.

			£	Skill Factor	£
Alteration work					
Hack off or take up					
wall plaster	1 m²	1:00	–	3	3.50
lath and plaster ceiling	1 m²	1:00	–	3	3.50
plasterboard ceiling	1 m²	1:00	–	3	3.50
wall tiling	1 m²	1:00	–	3	4.00
quarry tile floor	1 m²	1:00	–	3	3.50
Plastering					
One backing coat and one coat finishing plaster to walls	1 m²	1:40	2.70	8	12.00
One coat finishing plaster to plasterboard walls	1 m²	1:00	1.80	8	6.00
One coat finishing plaster to plasterboard ceilings	1 m²	1:20	1.80	8	7.00
Plasterboard 9.5 mm thick to walls	1 m²	1:00	2.90	6	8.00
Plasterboard 9.5 mm thick to ceilings	1 m²	1:20	2.90	6	9.00
Floor tiling					
Cement and sand floor screed trowelled smooth					
25 mm thick	1 m²	0:50	2.90	7	8.00
38 mm thick	1 m²	1:00	3.80	7	9.00
50 mm thick	1 m²	1:10	4.70	7	10.00

			$£$	Skill Factor	$£$
Red quarry tiles laid on screed					
150×150×12.5 mm	1 m²	1:30	20.00	7	40.00
200×200×12.5 mm	1 m²	1:00	19.00	7	37.00
Brown quarry tiles laid on screed					
150×150×12.5 mm	1 m²	1:30	26.00	7	47.00
200×200×12.5 mm	1 m²	1:00	25.00	7	44.00
Vinyl floor tiles size 300×300×2.5 mm thick, fixed with adhesive	1 m²	0:50	8.00	5	15.00
Vinyl floor sheeting 2.5 mm thick, fixed with adhesive	1 m²	0:50	9.00	5	18.00
Linoleum flooring 3.2 mm thick, fixed with adhesive	1 m²	0:50	10.00	5	20.00
Wall tiling					
White glazed ceramic wall tiling, fixed with adhesive and pointing with white grout					
108×108×4 mm	1 m²	1:00	18.00	7	30.00
152×152×5.6 mm	1 m²	0:50	17.00	7	28.00
203×108×6.5 mm	1 m²	0:50	16.00	7	27.00
Coloured glazed ceramic wall tiling, fixed with adhesive and pointing with white grout					
108×108×4 mm	1 m²	1:00	22.00	7	35.00
152×152×5.6 mm	1 m²	0:50	21.00	7	33.00
203×108×6.5 mm	1 m²	0:50	20.00	7	37.00

PLUMBING

A new technology in plumbing materials over recent years, such as light-weight materials and push-fit joints, has brought a wide range of activities that can now be carried out by DIY enthusiasts.

> **These prices may need adjusting to suit the price levels in your area.**
> See page xi in the introduction on how to adapt them
> for your particular part of the country

				Skill Factor	

Alteration work

Remove existing rainwater gutters and fittings, fix new length of gutter

cast iron, diameter

76 mm	1 no	1:30	17.00	5	35.00
115 mm	1 no	1:45	20.00	5	40.00

PVC-U, diameter

75 mm	1 no	1:00	7.00	5	25.00
110 mm	1 no	1:30	9.00	5	35.00

Remove rainwater gutter fittings, fit new

cast iron, 76 mm

angle	1 no	0:30	8.00	5	14.00
stop end outlet	1 no	0:25	4.00	5	10.00

cast iron, 115 mm

angle	1 no	0:45	9.00	5	18.00
stop end outlet	1 no	0:35	5.00	5	12.00

PVC-U, 75 mm

angle	1 no	0:20	4.00	5	6.00
outlet	1 no	0:20	4.00	5	6.00
stop end	1 no	0:15	2.00	5	3.00

PVC-U, 110 mm

angle	1 no	1:00	5.00	5	7.00
outlet	1 no	1:00	5.00	5	7.00
stop end	1 no	1:00	3.00	5	4.00

Remove existing rainwater pipe and fittings, fix new length of pipe

cast iron, diameter

75 mm	1 no	1:30	35.00	5	55.00
100 mm	1 no	1:45	48.00	5	70.00

PVC-U, diameter

68 mm	1 no	1:00	12.00	5	25.00
110 mm	1 no	1:30	20.00	5	35.00

Remove rainwater pipe fittings, fit new

cast iron, 75 mm

bend	1 no	0:30	19.00	5	30.00
shoe	1 no	0:25	15.00	5	25.00

cast iron, 100 mm

bend	1 no	0:30	15.00	5	25.00
shoe	1 no	0:25	20.00	5	30.00

				Skill Factor	

PVC-U, 68 mm					
bend	1 no	0:30	5.00	5	20.00
shoe	1 no	0:25	3.00	5	18.00
PVC-U, 110 mm					
bend	1 no	0:30	11.00	5	28.00
shoe	1 no	0:25	8.00	5	25.00
Cut out 500 mm length of copper pipe, fix new length with brass compression connections to existing ends					
15 mm	1 no	1:00	6.00	7	24.00
22 mm	1 no	1:10	12.00	7	30.00
28 mm	1 no	1:20	20.00	7	40.00
Take off existing radiator valve and replace with new, including bleeding down and draining system					
single standard-type valves, whole system	1 no	1:00	8.00	7	25.00
9 valves	1 no	12:00	72.00	7	240.00
Take out existing galvanised steel water storage tank, fix new plastic tank complete with ball valve, lid and insulation					
18 litres tank	1 no	6:00	18.00	7	110.00
68 litres tank	1 no	7:00	36.00	7	140.00
154 litres tank	1 no	8:00	44.00	7	180.00
Take out existing cast iron bath including trap and taps, cut back pipework as necessary and fix new acrylic reinforced bath complete with trap, bath panels, taps and shower handset and connect to existing pipework	1 no	8:00	300.00	7	450.00
Take out existing wash basin including trap and taps, cut back pipework as necessary and fix new pedestal-mounted wash basin complete with trap and taps and connect to existing pipework	1 no	8:00	110.00	7	250.00

	📏	🕐	£	Skill Factor	£
Take out existing high-level WC, cut back pipework as necessary and fix new low-level WC suite complete and connect to existing pipework	1 no	10:00	200.00	7	350.00

New work

Copper pipe with pre-soldered capillary joints and fittings

15 mm diameter	1 m	0:30	1.60	7	4.50
made bend	1 no	0:20	–	7	2.50
elbow	1 no	0:20	0.80	7	3.50
tee	1 no	0:25	1.20	7	4.00
tap connector	1 no	0:50	2.00	7	8.50
22 mm diameter	1 m	0:35	1.60	7	4.50
made bend	1 no	0:25	–	7	3.00
elbow	1 no	0:25	1.20	7	3.80
tee	1 no	0:30	2.80	7	4.00
tap connector	1 no	0:60	2.40	7	8.50
28 mm diameter	1 m	0:40	1.60	7	4.50
made bend	1 no	0:30	–	7	3.50
elbow	1 no	0:30	1.40	7	5.20
tee	1 no	0:35	4.20	7	4.00

Galvanised steel cold water storage cistern with cover, capacity

18 litres	1 no	4:00	50.00	7	90.00
36 litres	1 no	4:00	55.00	7	100.00
54 litres	1 no	4:30	60.00	7	110.00
68 litres	1 no	4:30	65.00	7	120.00

Plastic cold water storage cistern with cover, capacity

18 litres	1 no	4:00	20.00	7	60.00
68 litres	1 no	4:30	35.00	7	90.00
91 litres	1 no	5:00	45.00	7	110.00

Hot water copper direct cylinders

98 litres	1 no	5:00	80.00	7	140.00
120 litres	1 no	5:00	90.00	7	150.00
148 litres	1 no	5:00	100.00	7	165.00
166 litres	1 no	5:00	120.00	7	180.00

				Skill Factor	

Hot water copper indirect
cylinders

96 litres	1 no	5:00	95.00	7	160.00
114 litres	1 no	5:00	110.00	7	175.00
140 litres	1 no	5:00	125.00	7	190.00
162 litres	1 no	5:00	160.00	7	210.00

Insulation

Preformed pipe lagging,
fire-retardant foam 13 mm thick
for pipe size

15 mm	1 m	0:05	2.20	4	3.50
22 mm	1 m	0:06	2.90	4	4.20
28 mm	1 m	0:07	3.30	4	5.00

Expanded polystryrene lagging
jacket set to bottom and sides
of galvanised steel cisterns, size

445×305×300 mm	1 no	0:30	8.00	4	15.00
495×368×362 mm	1 no	0:35	9.00	4	18.00
630×450×420 mm	1 no	0:40	10.00	4	20.00

PVC-U insulating jacket 80 mm
thick, filled with expanded
polystyrene securing with fixing
bands to hot water cylinder, size

400×1050 mm	1 no	0:15	10.00	4	15.00
450×900 mm	1 no	0:15	12.00	4	18.00
400×1200 mm	1 no	0:15	14.00	4	20.00

32 mm diameter MPVC-U waste

pipe with push-fit joints	1 m	0:40	3.00	7	5.00
bend	1 no	0:40	2.00	7	5.00
tee	1 no	0:40	3.00	7	6.00
P trap	1 no	0:40	5.00	7	8.00
S trap	1 no	0:40	6.00	7	10.00

40 mm diameter MPVC-U waste

pipe with push-fit joints	1 m	0:45	3.50	7	5.00
bend	1 no	0:40	2.00	7	5.00
tee	1 no	0:40	3.00	7	6.00
P trap	1 no	0:40	5.00	7	8.00
S trap	1 no	0:40	6.00	7	10.00

				Skill Factor	
Acrylic reinforced bath, white, size 1700×700 mm complete with chromium plated grips, 40 mm waste fitting, overflow with chain and plastic plug	1 no	6.00	160.00	7	230.00
Porcelain enamelled heavy gauge bath, white, size 1500×700 mm complete with chromium plated grips, 40 mm waste fitting, overflow with chain and plastic plug	1 no	6.00	200.00	7	280.00
Bath panels, enamelled hardboard					
end panel	1 no	0.30	6.00	5	10.00
side panel	1 no	0.30	9.00	5	15.00
Bath panels, moulded acrylic					
end panel	1 no	0.30	8.00	5	12.00
side panel	1 no	0.30	15.00	5	18.00
Aluminium angle strip					
25×25×560 mm	1 no	0.30	4.00	5	8.00
Wash basin, vitreous china, size 510×410 mm, complete					
white	1 no	3.00	75.00	7	130.00
coloured	1 no	3.00	85.00	7	140.00
Wash basin, vitreous china, size 610×470 mm complete					
white	1 no	3.20	80.00	7	135.00
coloured	1 no	3.20	90.00	7	145.00
Stainless steel single sink and drainer complete, size					
1000×500 mm	1 no	2.00	80.00	7	110.00
1000×600 mm	1 no	2.00	95.00	7	115.00
1200×600 mm	1 no	2.00	90.00	7	120.00
Belfast pattern white fireclay sink, size					
450×380×205 mm	1 no	2.30	100.00	7	160.00
610×455×255 mm	1 no	2.45	150.00	7	220.00
760×455×255 mm	1 no	3.00	200.00	7	275.00
Vitreous china low-level WC suite complete	1 no	3.00	210.00	7	300.00
Vitreous china low-level close-coupled WC suite complete	1 no	3.00	240.00	7	350.00

			Skill Factor	
Chromium-plated thermostatic exposed shower valve with flexible hose, slide rail and spray set	1 no	1.30	190.00	7	230.00
Instant electric 9.5 kw shower with flexible hose, slide rail and spray set	1 no	1.30	140.00	7	210.00
Shower cubicle size 788×842× 2115 mm in anodised aluminium frame and safety glass	1 no	2.00	400.00	7	470.00
White glazed fireclay shower tray size 900×900×180 mm	1 no	2.00	160.00	7	220.00
White acrylic fireclay shower tray size 900×900×180 mm	1 no	1.30	95.00	7	130.00
Single-panel radiator fixed to wall with concealed brackets					
450 mm high, length					
500 mm	1 no	1.00	22.00	7	40.00
1000 mm	1 no	1.10	45.00	7	65.00
1600 mm	1 no	1.20	70.00	7	90.00
2000 mm	1 no	1.30	100.00	7	130.00
600 mm high, length					
500 mm	1 no	1.10	30.00	7	50.00
1000 mm	1 no	1.20	55.00	7	75.00
1600 mm	1 no	1.30	90.00	7	120.00
2000 mm	1 no	1.40	130.00	7	170.00
700 mm high, length					
500 mm	1 no	1.20	35.00	7	65.00
1000 mm	1 no	1.30	70.00	7	100.00
1600 mm	1 no	1.40	120.00	7	160.00
2000 mm	1 no	1.50	150.00	7	200.00
Single-panel single convector radiator fixed to wall with concealed brackets					
450 mm high, length					
500 mm	1 no	1.00	22.00	7	40.00
1000 mm	1 no	1.10	45.00	7	65.00
1600 mm	1 no	1.20	70.00	7	90.00
2000 mm	1 no	1.30	130.00	7	160.00
600 mm high, length					
500 mm	1 no	1.10	30.00	7	50.00
1000 mm	1 no	1.20	50.00	7	70.00

				Skill Factor	
1600 mm	1 no	1.30	90.00	7	120.00
2000 mm	1 no	1.40	170.00	7	210.00
700 mm high, length					
500 mm	1 no	1.10	35.00	7	65.00
1000 mm	1 no	1.20	70.00	7	100.00
1600 mm	1 no	1.30	160.00	7	200.00
2000 mm	1 no	1.40	195.00	7	245.00
Double-panel single convector radiator fixed to wall with concealed brackets					
450 mm high, length					
500 mm	1 no	1.10	30.00	7	60.00
1000 mm	1 no	1.20	60.00	7	90.00
1600 mm	1 no	1.30	110.00	7	150.00
2000 mm	1 no	1.40	190.00	7	240.00
600 mm high, length					
500 mm	1 no	1.20	40.00	7	70.00
1000 mm	1 no	1.30	80.00	7	110.00
1600 mm	1 no	1.40	140.00	7	180.00
2000 mm	1 no	1.50	240.00	7	290.00
700 mm high, length					
500 mm	1 no	1.30	45.00	7	75.00
1000 mm	1 no	1.40	90.00	7	120.00
1600 mm	1 no	1.50	215.00	7	255.00
2000 mm	1 no	2.00	270.00	7	320.00
Double-panel double convector radiator fixed to wall with concealed brackets					
450 mm high, length					
500 mm	1 no	1.20	40.00	7	70.00
1000 mm	1 no	1.30	80.00	7	110.00
1600 mm	1 no	1.40	150.00	7	190.00
2000 mm	1 no	1.50	240.00	7	290.00
600 mm high, length					
500 mm	1 no	1.30	50.00	7	80.00
1000 mm	1 no	1.40	100.00	7	130.00
1600 mm	1 no	1.50	190.00	7	230.00
2000 mm	1 no	2.00	310.00	7	360.00
2400 mm	1 no	2.10	370.00	7	420.00
700 mm high, length					
500 mm	1 no	1.40	60.00	7	95.00
1000 mm	1 no	1.50	115.00	7	145.00

				Skill Factor	
1600 mm	1 no	2.00	280.00	7	320.00
2000 mm	1 no	2.10	350.00	7	400.00
2400 mm	1 no	2.20	420.00	7	470.00

GLAZING

Most households suffer from a broken window from time to time and the following costs should help in deciding whether to do it yourself or call in a contractor. Information is also given on the cost of factory-made double glazing units that match the recent changes to the building regulations for insulation.

These changes require the insulation properties of double glazing to be subject to low emissivity or Low E. The effect of this type of glass enables it to allow the heat from the sun to pass through the glass but reflects any internally generated (from fires or radiators) to be reflected back into the room.

				Skill Factor	
Hack out broken glass from wood or metal windows and re-glaze with 4 mm thick clear glass in putty					
225×450 mm	1 no	1:00	6.00	5	18.00
450×450 mm	1 no	1:10	10.00	5	20.00
750×600 mm	1 no	1:20	14.00	5	26.00
900×750 mm	1 no	1:30	18.00	5	30.00
Hack out broken glass from wood or metal windows and re-glaze with 4 mm thick, clear glass in wooden beads previously laid aside					
225×450 mm	1 no	1:10	6.00	5	18.00
450×450 mm	1 no	1:20	10.00	5	20.00
750×600 mm	1 no	1:30	14.00	5	26.00
900×750 mm	1 no	1:40	18.00	5	30.00

	📏	🕐	£	Skill Factor	⚡
Hack out broken glass from wood or metal windows and re-glaze with 4 mm thick clear glass in new beads					
225×450 mm	1 no	1:00	7.00	5	20.00
450×450 mm	1 no	1:10	11.00	5	22.00
750×600 mm	1 no	1:20	16.00	5	28.00
900×750 mm	1 no	1:30	20.00	5	32.00
Hack out broken glass from wood or metal windows and re-glaze with 6 mm thick obscure glass in putty					
225×450 mm	1 no	1:00	9.00	5	21.00
450×450 mm	1 no	1:10	15.00	5	27.00
750×600 mm	1 no	1:20	21.00	5	33.00
900×750 mm	1 no	1:30	27.00	5	38.00
Hack out broken glass from wood or metal windows and re-glaze with 6 mm thick obscure glass in wooden beads previously laid aside					
225×450 mm	1 no	1:10	9.00	5	21.00
450×450 mm	1 no	1:20	15.00	5	27.00
750×600 mm	1 no	1:30	21.00	5	33.00
900×750 mm	1 no	1:40	27.00	5	38.00
Hack out broken glass from wood or metal windows and re-glaze with 6 mm thick obscure glass in new beads					
225×450 mm	1 no	1:00	11.00	5	21.00
450×450 mm	1 no	1:10	16.00	5	27.00
750×600 mm	1 no	1:20	24.00	5	33.00
900×750 mm	1 no	1:30	30.00	5	38.00
Hermetically sealed double glazing units in clear glass 24 mm overall thickness 580 mm wide					
580×400 mm	1 no	1:00	20.00	5	40.00
580×550 mm	1 no	1:10	22.00	5	42.00
580×700 mm	1 no	1:20	26.00	5	46.00
580×850 mm	1 no	1:30	33.00	5	53.00

				Skill Factor	

580×1000 mm	1 no	1:40	40.00	5	60.00
580×1200 mm	1 no	1:50	45.00	5	65.00

Hermetically sealed double
glazing units in clear glass
24 mm overall thickness
860 mm wide

860×400 mm	1 no	1:00	20.00	5	50.00
860×550 mm	1 no	1:10	24.00	5	54.00
860×700 mm	1 no	1:20	32.00	5	62.00
860×850 mm	1 no	1:30	40.00	5	70.00
860×1000 mm	1 no	1:40	50.00	5	80.00
860×1200 mm	1 no	1:50	60.00	5	90.00

Hermetically sealed double
glazing units in clear glass
24 mm overall thickness
1100 mm wide

1100×400 mm	1 no	1:00	30.00	5	70.00
1100×550 mm	1 no	1:10	44.00	5	84.00
1100×700 mm	1 no	1:20	58.00	5	98.00
1100×850 mm	1 no	1:30	70.00	5	110.00
1100×1000 mm	1 no	1:40	84.00	5	124.00
1100×1200 mm	1 no	1:50	98.00	5	138.00

Hermetically sealed double
glazing units in obscure glass
24 mm overall thickness,
580 mm wide

580×400 mm	1 no	1:00	24.00	5	44.00
580×550 mm	1 no	1:10	26.00	5	46.00
580×700 mm	1 no	1:20	30.00	5	50.00
580×850 mm	1 no	1:30	38.00	5	58.00
580×1000 mm	1 no	1:40	46.00	5	66.00
580×1200 mm	1 no	1:50	52.00	5	72.00

Hermetically sealed double
glazing units in obscure glass
24 mm overall thickness,
860 mm wide

860×400 mm	1 no	1:00	24.00	5	54.00
860×550 mm	1 no	1:10	28.00	5	58.00
860×700 mm	1 no	1:20	36.00	5	66.00
860×850 mm	1 no	1:30	46.00	5	76.00

	⊞	🕐	£	Skill Factor	£
860 × 1000 mm	1 no	1:40	54.00	5	84.00
860 × 1200 mm	1 no	1:50	60.00	5	90.00
Hermetically sealed double glazing units in obscure glass 24 mm overall thickness, 1100 mm wide					
1100 × 400 mm	1 no	1:00	36.00	5	86.00
1100 × 550 mm	1 no	1:10	50.00	5	100.00
1100 × 700 mm	1 no	1:20	64.00	5	114.00
1100 × 850 mm	1 no	1:30	76.00	5	124.00
1100 × 1000 mm	1 no	1:40	90.00	5	140.00
1100 × 1200 mm	1 no	1:50	104.00	5	154.00
Hermetically sealed double glazing units in clear glass with diamond or rectangular leaded panes, 24 mm overall thickness, 580 mm wide					
580 × 400 mm	1 no	1:00	50.00	5	70.00
580 × 550 mm	1 no	1:10	56.00	5	76.00
580 × 700 mm	1 no	1:20	76.00	5	96.00
580 × 850 mm	1 no	1:30	84.00	5	104.00
580 × 1000 mm	1 no	1:40	100.00	5	170.00
580 × 1200 mm	1 no	1:50	115.00	5	185.00
Hermetically sealed double glazing units in clear glass with diamond or rectangular leaded panes, 24 mm overall thickness, 860 mm wide					
860 × 400 mm	1 no	1:00	50.00	5	80.00
860 × 550 mm	1 no	1:10	60.00	5	90.00
860 × 700 mm	1 no	1:20	84.00	5	114.00
860 × 850 mm	1 no	1:30	100.00	5	130.00
860 × 1000 mm	1 no	1:40	125.00	5	155.00
860 × 1200 mm	1 no	1:50	150.00	5	180.00
Hermetically sealed double glazing units in clear glass with diamond or rectangular leaded panes, 24 mm overall thickness, 1100 mm wide					
1100 × 400 mm	1 no	1:00	75.00	5	115.00
1100 × 550 mm	1 no	1:10	110.00	5	150.00

				Skill Factor	
1100×700 mm	1 no	1:20	140.00	5	180.00
1100×850 mm	1 no	1:30	175.00	5	215.00
1100×1000 mm	1 no	1:40	210.00	5	250.00
1100×1200 mm	1 no	1:50	246.00	5	286.00

ELECTRICAL WORK

Only the most basic electrical tasks should be carried out by a DIYer. The paradox is that a high proportion of an electrician's time is spent in non-technical work – drilling holes, lifting floorboards and the like – but making the connections requires a level of skill that is better left to an expert.

The prices quoted below are those that would be charged by an electrical firm for a reasonable amount of work in a normal domestic situation.

	Unit	Electrician's charges £
Form spur from existing point for a distance not exceeding 3 metres and provide		
single power point	1 no	45.00
twin power point	1 no	55.00
light switch, 1 gang	1 no	30.00
light switch, 2 gang	1 no	35.00
pendant	1 no	40.00
wall light	1 no	50.00
lighting including flex and lamp holder	1 no	50.00
shaver point	1 no	45.00
cooker control unit	1 no	80.00
Form spur from existing point for a distance not exceeding 6 metres and provide		
single power point	1 no	50.00
twin power point	1 no	60.00
light switch, 1 gang	1 no	35.00
light switch, 2 gang	1 no	40.00
pendant	1 no	45.00
wall light	1 no	55.00
lighting including flex and lamp holder	1 no	55.00

	Unit	Electrician's charges £
shaver point	1 no	50.00
cooker control unit	1 no	85.00
Supply and fix single tube fluorescent light fitting adjacent to existing ceiling point, length		
600 mm	1 no	22.00
1200 mm	1 no	30.00
1500 mm	1 no	32.00
1800 mm	1 no	39.00
Supply and fix twin tube fluorescent light fitting adjacent to existing ceiling point, length		
600 mm	1 no	30.00
1200 mm	1 no	38.00
1500 mm	1 no	44.00
1800 mm	1 no	50.00

DECORATING AND PAPERHANGING

This is one of the most popular of DIY activities and most people believe that they can make a fair job of painting and papering a room although the standard of finish can vary widely. The section is presented in two parts; first, the times and costs per unit for a wide range of items, second, using this data for the times and costs of typical size rooms.

				Skill Factor	
Preparation					
Scrape off wallpaper from existing					
walls	1 m²	0:30	–	1	1.50
ceilings	1 m²	0:40	–	1	2.90
Wash down painted softwood or metal surfaces and rub down to receive new paint					
windows, doors and general surfaces	1 m²	0:20	–	1	2.20
frames and rails less than 200 mm	1 m	0:10	–	1	1.50

				Skill Factor	
Burn off paint from softwood or metal surfaces and rub down to receive new paint					
windows, doors and general surfaces	1 m²	1:00	–	1	5.50
frames and rails less than 200 mm	1 m	0:30	–	1	2.50
New painting work – standard items					
One coat of emulsion paint					
plastered walls	1 m²	0:10	0.50	5	2.20
plastered ceilings	1 m²	0:15	0.50	5	2.50
Two coats of emulsion paint					
plastered walls	1 m²	0:20	1.00	5	3.20
plastered ceilings	1 m²	0:30	1.00	5	3.50
One coat primer to wood					
general surfaces	1 m²	0:20	0.50	5	2.40
skirtings less than 150 mm high	1 m	0:04	0.08	5	0.80
frames and rails less than 200 mm	1 m	0:06	0.10	5	0.90
One coat primer to metal					
general surfaces	1 m²	0:20	0.60	5	2.80
skirtings less than 150 mm high	1 m	0:04	0.09	5	0.80
frames and rails less than 200 mm	1 m	0:06	0.12	5	1.00
One coat undercoat to wood or metal surfaces					
general surfaces	1 m²	0:20	0.60	5	2.80
skirtings less than 150 mm high	1 m	0:04	0.08	5	0.80
frames and rails less than 200 mm	1 m	0:06	0.12	5	1.00
One coat gloss to wood or metal surfaces					
general surfaces	1 m²	0:20	0.60	5	2.80
skirtings less than 150 mm high	1 m	0:04	0.08	5	0.80
frames and rails less than 200 mm	1 m	0:06	0.12	5	1.00
One coat primer, one coat undercoat and one coat gloss to wood surfaces					
general surfaces	1 m²	1:00	1.70	5	8.00
skirtings less than 150 mm high	1 m	0:12	0.25	5	2.40
frames and rails less than 200 mm	1 m	0:18	0.34	5	2.90
One coat primer, one coat undercoat and one coat gloss to metal surfaces					
general surfaces	1 m²	1:00	1.80	5	8.20
skirtings less than 150 mm high	1 m	0:12	0.25	5	2.40
frames and rails less than 200 mm	1 m	0:18	0.40	5	3.00

New wall papering

The basic price (BP) is the retail cost of a roll of wallpaper. The costs in the 'Material cost' column includes an allowance for wallpaper paste and filling in any cracks in the plaster.

				Skill Factor	
Supply and hang wallpaper to plastered walls					
lining paper					
(BP £1.00)	1 m²	0:20	0.20	6	1.90
(BP £1.50)	1 m²	0:20	0.30	6	2.00
woodchip paper					
(BP £1.50)	1 m²	0:20	0.30	6	2.00
(BP £2.00)	1 m²	0:20	0.40	6	2.10
(BP £2.50)	1 m²	0:20	0.50	6	2.20
standard patterned paper					
(BP £4.50)	1 m²	0:25	0.88	6	2.00
(BP £5.50)	1 m²	0:25	1.05	6	2.10
(BP £7.00)	1 m²	0:25	1.35	6	2.20
vinyl-surfaced paper					
(BP £6.00)	1 m²	0:25	1.15	6	2.15
(BP £7.00)	1 m²	0:25	1.35	6	2.20
(BP £8.00)	1 m²	0:25	1.55	6	2.40
flock paper					
(BP £8.00)	1 m²	0:30	1.55	6	3.25
(BP £9.00)	1 m²	0:30	1.70	6	3.40
(BP £10.00)	1 m²	0:30	1.90	6	3.60
Supply and hang wallpaper to plastered ceilings					
lining paper					
(BP £1.00)	1 m²	0:30	0.20	6	2.40
(BP £1.50)	1 m²	0:30	0.30	6	2.50
woodchip paper					
(BP £1.50)	1 m²	0:30	0.30	6	2.50
(BP £2.00)	1 m²	0:30	0.40	6	2.60
(BP £2.50)	1 m²	0:30	0.50	6	2.70

The next section gives the net areas and lengths of surfaces to be painted or papered for rooms in a typical house. The following assumptions have been made.

Living room	1 door, 1 fireplace, 1 large window
Dining room	2 doors, 1 large window
Bedroom	1 door, 1 average window
Kitchen	2 doors, fittings below dado level, 1 large window
WC	1 door, 1 small window
Bathroom	1 door, 1 average window, 1 cupboard
Ceiling heights	Ground floor 2.4 metres
	First floor 2.2 metres.

Typical quantities for painting and papering.

	Ceiling	Walls	Window	Door	Skirting	Frames
Living room						
3.6×3.0 m	11 m²	24 m²	3 m²	2 m²	10 m	8 m
3.6×3.6 m	13 m²	26 m²	3 m²	2 m²	11 m	8 m
4.2×3.0 m	13 m²	26 m²	3 m²	2 m²	11 m	8 m
4.2×3.6 m	15 m²	29 m²	3 m²	2 m²	13 m	8 m
4.2×4.2 m	18 m²	32 m²	3 m²	2 m²	14 m	8 m
4.8×3.0 m	15 m²	29 m²	3 m²	2 m²	13 m	8 m
4.8×3.6 m	17 m²	32 m²	3 m²	2 m²	14 m	8 m
4.8×4.2 m	20 m²	34 m²	3 m²	2 m²	15 m	8 m
4.8×4.8 m	23 m²	37 m²	3 m²	2 m²	16 m	8 m
Dining room						
3.0×3.0 m	9 m²	20 m²	3 m²	4 m²	10 m	13 m
3.6×3.0 m	11 m²	24 m²	3 m²	4 m²	11 m	13 m
3.6×3.6 m	13 m²	26 m²	3 m²	4 m²	13 m	13 m
4.2×3.0 m	13 m²	26 m²	3 m²	4 m²	13 m	13 m
4.2×3.6 m	15 m²	29 m²	3 m²	4 m²	15 m	13 m
4.2×4.2 m	18 m²	32 m²	3 m²	4 m²	16 m	13 m
4.8×3.0 m	15 m²	29 m²	3 m²	4 m²	15 m	13 m
4.8×3.6 m	17 m²	32 m²	3 m²	4 m²	16 m	13 m
4.8×4.2 m	20 m²	34 m²	3 m²	4 m²	17 m	13 m
Bedroom						
2.4×2.4 m	6 m²	17 m²	2 m²	2 m²	9 m	7 m
2.4×3.0 m	7 m²	20 m²	2 m²	2 m²	10 m	7 m
2.4×3.6 m	9 m²	22 m²	2 m²	2 m²	11 m	7 m
3.0×3.0 m	9 m²	22 m²	2 m²	2 m²	11 m	7 m
3.0×3.6 m	11 m²	25 m²	2 m²	2 m²	12 m	7 m
3.6×3.6 m	13 m²	27 m²	2 m²	2 m²	13 m	7 m
3.6×4.2 m	15 m²	30 m²	2 m²	2 m²	15 m	7 m
3.6×4.8 m	17 m²	32 m²	2 m²	2 m²	16 m	7 m
4.2×4.2 m	18 m²	32 m²	2 m²	2 m²	16 m	7 m
4.2×4.8 m	20 m²	35 m²	2 m²	2 m²	17 m	7 m

	Ceiling	Walls	Window	Door	Skirting	Frames
Kitchen						
1.8×3.0 m	6 m²	7 m²	3 m²	4 m²	3 m	13 m
1.8×3.6 m	7 m²	10 m²	3 m²	4 m²	4 m	13 m
2.4×3.0 m	7 m²	10 m²	3 m²	4 m²	4 m	13 m
2.4×3.6 m	9 m²	11 m²	3 m²	4 m²	5 m	13 m
3.0×3.0 m	9 m²	11 m²	3 m²	4 m²	5 m	13 m
3.0×3.6 m	11 m²	11 m²	3 m²	4 m²	6 m	13 m
3.0×4.2 m	13 m²	11 m²	3 m²	4 m²	6 m	13 m
3.6×3.6 m	13 m²	13 m²	3 m²	4 m²	6 m	13 m
3.6×4.2 m	15 m²	15 m²	3 m²	4 m²	7 m	13 m
WC						
1.0×1.5 m	2 m²	8 m²	1 m²	2 m²	4 m	6 m
1.2×1.5 m	2 m²	9 m²	1 m²	2 m²	4 m	6 m
1.2×1.8 m	2 m²	10 m²	1 m²	2 m²	5 m	6 m
1.3×1.5 m	2 m²	9 m²	1 m²	2 m²	5 m	6 m
1.3×1.8 m	2 m²	10 m²	1 m²	2 m²	5 m	6 m
Bathroom size						
1.8×2.4 m	4 m²	23 m²	1 m²	6 m²	9 m	14 m
1.8×3.0 m	5 m²	25 m²	1 m²	6 m²	11 m	14 m
2.1×2.4 m	5 m²	23 m²	1 m²	6 m²	10 m	14 m
2.4×2.4 m	6 m²	25 m²	1 m²	6 m²	10 m	14 m
2.4×3.0 m	7 m²	28 m²	1 m²	6 m²	11 m	14 m
2.4×3.6 m	9 m²	30 m²	1 m²	6 m²	13 m	14 m

These quantities can now be combined with the costs set out in standard item to produce contractor's charges for each element of individual rooms. It is assumed that the walls will be papered with vinyl-surfaced wallpaper at a cost of £6.00 per roll. All figures have been rounded off to the nearest pound.

	Ceiling £	Walls £	Window £	Door £	Skirting £	Frames £
Living room						
3.6×3.0 m	39	52	24	16	24	24
3.6×3.6 m	46	56	24	16	27	24
4.2×3.0 m	46	56	24	16	27	24
4.2×3.6 m	53	63	24	16	31	24
4.2×4.2 m	63	69	24	16	34	24
4.8×3.0 m	53	63	24	16	31	24
4.8×3.6 m	60	69	24	16	34	24
4.8×4.2 m	70	73	24	16	36	24
4.8×4.8 m	81	80	24	16	39	24

	Ceiling £	Walls £	Window £	Door £	Skirting £	Frames £
Dining room						
3.0×3.0 m	32	43	24	32	24	39
3.6×3.0 m	39	52	24	32	27	39
3.6×3.6 m	46	56	24	32	31	39
4.2×3.0 m	46	56	24	32	31	39
4.2×3.6 m	53	63	24	32	36	39
4.2×4.2 m	63	69	24	32	39	39
4.8×3.0 m	53	63	24	32	36	39
4.8×3.6 m	60	69	24	32	39	39
4.8×4.2 m	70	73	24	32	41	39
Bedroom						
2.4×2.4 m	21	37	16	16	22	21
2.4×3.0 m	25	43	16	16	24	21
2.4×3.6 m	32	47	16	16	26	21
3.0×3.0 m	32	47	16	16	26	21
3.0×3.6 m	39	54	16	16	29	21
3.6×3.6 m	46	58	16	16	31	21
3.6×4.2 m	53	65	16	16	36	21
3.6×4.8 m	60	69	16	16	38	21
4.2×4.2 m	63	69	16	16	38	21
4.8×4.8 m	70	75	16	16	41	21
Bedroom						
1.8×2.4 m	21	37	16	16	22	21
1.8×3.6 m	25	22	24	32	10	39
2.4×3.0 m	25	22	24	32	10	39
2.4×3.6 m	32	24	24	32	12	39
3.0×3.0 m	32	24	24	32	12	39
3.0×3.6 m	39	24	24	32	14	39
3.0×4.2 m	46	24	24	32	14	39
3.6×3.6 m	46	28	24	32	14	39
3.6×4.2 m	53	32	24	32	17	39
WC						
1.0×1.5 m	7	17	8	16	10	18
1.2×1.5 m	7	19	8	16	10	18
1.2×1.8 m	7	22	8	16	12	18
1.3×1.5 m	7	19	8	16	12	18
1.3×1.8 m	7	22	8	16	12	18
Bathroom						
1.8×2.4 m	14	50	8	48	22	42
1.8×3.0 m	18	54	8	48	26	42
2.1×2.4 m	18	50	8	48	24	42
2.4×2.4 m	21	54	8	48	24	42
2.4×3.0 m	25	60	8	48	26	42
2.4×3.6 m	32	65	8	48	31	42

Here is an example of decorating a bedroom of size 3.6×4.2 m. The work consists of:

- scraping off the existing wallpaper and hanging new vinyl-surfaced paper costing £8.00 per roll;
- wash down painted softwood or metal surfaces and rub down to receive new paint;
- apply one undercoat and one coat gloss to metal and wood surfaces;
- two coats emulsion paint to the ceiling.

				Skill Factor	
Scrape off existing wallpaper	30 m²	15:00	–	1	
Hang vinyl-surfaced wallpaper	30 m²	20:00	46.50	6	
Wash down existing paintwork and rub down to receive new					
general surfaces	4 m²	1:20	–	1	
frames and rails	22 m	3:40	–	1	
One coat undercoat and one coat gloss to					
general surfaces	4 m²	2:40	–	5	
skirtings	15 m	3:00	2.40	5	
frames	7 m	1:24	2.03	5	
Two coats emulsion paint to ceilings	15 m²	7:30	15.00	5	
		54.34	65.93		550

You can see from this example that it should take about 55 hours to redecorate your bedroom and the materials cost is £65.93. On the other hand, the cost of employing a painter would be about £550, so you could save over £400 by doing the work yourself.

These prices may need adjusting to suit the price levels in your area.
See page xi in the introduction on how to adapt them
for your particular part of the country

	▤	🕐	£	Skill Factor	♠
External painting					
Wash down painted softwood or metal surfaces and rub down to receive new paint					
windows, doors and general surfaces	1 m²	0:25	–	1	2.50
frames and rails less than 200 mm	1 m	0:12	–	1	1.75
Burn off paint from softwood or metal surfaces and rub down to receive new paint					
windows, doors and general surfaces	1 m²	1:10	–	1	6.00
frames and rails less than 200 mm	1 m	0:35	–	1	2.75
One coat undercoat and one coat gloss to wood surfaces					
general surfaces	1 m²	1:10	1.70	5	8.50
skirtings less than 150 mm high	1 m	0:15	0.25	5	2.60
frames and rails less than 200 mm	1 m	0:20	0.34	5	3.20
One coat undercoat and one coat gloss to metal surfaces					
general surfaces	1 m²	1:10	1.80	5	8.70
skirtings less than 150 mm high	1 m	0:15	0.25	5	2.80
frames and rails less than 200 mm	1 m	0:20	0.40	5	3.40
Two coats cement paint					
brick or block walls	1 m²	0:45	2.00	4	6.00
cement rendering	1 m²	0:40	1.80	4	5.50
Two coats creosote on wood surfaces	1 m²	0:20	0.30	4	2.00
Two coats wood preservative on wood surfaces	1 m²	0:18	1.20	4	3.00

PATHS AND EDGINGS

The figures in this section are presented in two parts. First, preparatory work needed for all types of paving and second, costs and times for different types of paving materials. It is assumed that all the excavated material can be disposed off in the garden. See *Disposal of Material* on page 102 if it is necessary to remove the earth from the site.

		📏	🕐	💷	Skill Factor	💲

Preparatory work

Excavate to remove vegetable soil 150 mm and spread in garden	1 m²	0:30	–	2	
Bed of sand 100 mm thick	1 m²	0:15	0.30	2	
per square metre		0:45	0.30		4.50

Precast concrete slabs

Natural colour slabs, size					
450×450×50 mm	1 m²	0:45	12.00	4	18.00
600×600×50 mm	1 m²	0:40	8.50	4	15.00
750×600×50 mm	1 m²	0:35	8.00	4	14.50
900×600×50 mm	1 m²	0:30	7.50	4	13.00
Coloured slabs, size					
450×450×50 mm	1 m²	0:45	13.00	4	19.00
600×600×50 mm	1 m²	0:40	9.50	4	16.00
750×600×50 mm	1 m²	0:35	9.00	4	15.50
900×600×50 mm	1 m²	0:30	8.50	4	14.00
Brick paving, £450 per thousand					
laid flat	1 m²	1:20	24.00	6	36.00
laid on edge	1 m²	1:45	34.00	6	48.00
herringbone pattern laid flat	1 m²	1:50	24.00	6	40.00
herringbone pattern laid on edge	1 m²	2:10	34.00	6	52.00
Brick paving, £550 per thousand					
laid flat	1 m²	1:20	30.00	6	42.00
laid on edge	1 m²	1:45	42.00	6	56.00
herringbone pattern laid flat	1 m²	1:50	30.00	6	46.00
herringbone pattern laid on edge	1 m²	2:10	42.00	6	60.00
Cobble paving, £100 per tonne					
laid to regular pattern	1 m²	4:30	25.00	6	75.00
laid to irregular pattern	1 m²	5:30	25.00	6	85.00
York stone paving, slab size					
600×600×50 mm	1 m²	0:50	90.00	5	105.00
900×600×50 mm	1 m²	0:40	95.00	5	110.00
Granite sett paving size 100×100×75 mm	1 m²	1:40	50.00	5	72.00

Edgings

Sometimes a paved area needs a small pin kerb or a brick-on-edge to mark the boundary with other types of textured finishes. In the costs below it is assumed that the concrete is mixed by hand and that the excavated material will be spread over the garden.

			£	Skill Factor	
Precast concrete edgings					
Excavate shallow trench size 300 × 150 mm for kerb foundation	1 m	0:10	–	2	
Concrete in foundation	1 m	0:20	3.60	2	
Precast concrete pin kerb size 150 × 50 mm	1 m	0:25	3.20	5	
per linear metre		0:55	7.20		12.00
Brick edgings					
Excavate shallow trench size 300 × 150 mm for kerb foundation	1 m	0:10	–	2	
Concrete in foundation	1 m	0:20	3.60	2	
Facing bricks (£250 per thousand) in brick edging	1 m	1:00	4.00	5	
per linear metre		1:30	7.60		16.00
Excavate shallow trench size 300 × 150 mm for kerb foundation	1 m	0:10	–	2	
Concrete in foundation	1 m	0:20	3.60	2	
Facing bricks (£350 per thousand) in brick edging	1 m	1:00	6.50	5	
per linear metre		1:30	9.10		18.00
Excavate shallow trench size 300 × 150 mm for kerb foundation	1 m	0:10	–	2	
Concrete in foundation	1 m	0:20	3.60	2	
Facing bricks (£450 per thousand) in brick edging	1 m	1:00	7.50	5	
per linear metre		1:30	11.10		20.00

FENCING

All fencing work is within the range of DIY enthusiasts and the erection of fencing can be a satisfying project. The figures set out below are based on the assumptions that the post holes are dug out by hand and that the concrete is ready mixed and delivered and deposited within 25 metres of its point of placing.

				Skill Factor	
Chainlink fencing, galvanised steel mesh on three strained line wires fixed to concrete posts at 3 m centres					
900 mm high	1 m	1:20	10.00	5	18.00
1200 mm high	1 m	1:40	12.00	5	22.00
1800 mm high	1 m	2:00	14.00	5	26.00
Chainlink fencing, galvanised steel mesh on three strained line wires fixed to galvanised steel posts at 3 m centres					
900 mm high	1 m	1:20	9.00	5	17.00
1200 mm high	1 m	1:40	11.00	5	20.00
1800 mm high	1 m	2:00	13.00	5	24.00
Strained wire fence including concrete posts at 3 m centres					
1.00 m high, 5 wires	1 m	0:40	6.00	5	12.00
1.20 m high, 6 wires	1 m	1:00	8.00	5	14.00
1.40 m high, 8 wires	1 m	1:20	10.00	5	16.00
Cleft chestnut fencing, pales set 75 mm apart, 75 mm diameter softwood posts at 2500 mm centres					
1.00 m high, 2 wires	1 m	0:40	6.00	5	12.00
1.20 m high, 2 wires	1 m	1:00	8.00	5	14.00
1.50 m high, 2 wires	1 m	1:20	10.00	5	16.00
1.80 m high, 3 wires	1 m	1:40	12.00	5	18.00
Close-boarded fencing consisting of 90×19 mm pales lapped 13 mm fixed to 75×38 mm horizontal rails on 75×75 mm timber posts at 3 m centres, height					
1.00 m	1 m	2:00	20.00	5	34.00
1.20 m	1 m	2:20	22.00	5	38.00
1.50 m	1 m	2:40	24.00	5	42.00

		⏱	£	Skill Factor	£
Close-boarded fencing consisting of 90×19 mm pales lapped 13 mm fixed to 75×38 mm horizontal rails on 75×75 mm concrete posts at 3 m centres, height					
1.00 m	1 m	2:00	14.00	5	26.00
1.20 m	1 m	2:20	16.00	5	30.00
1.50 m	1 m	2:40	18.00	5	34.00

PATIOS

Laying a patio can add a practical and attractive feature to your house. There are four main types of materials used in the construction of a patio – concrete or stone paving slabs, brick paving and granite setts. Slabs are quicker to lay than bricks and setts because of their size but this short-term benefit should not influence your decision on which material to use.

All these materials are durable and should last over 20 years without any deterioration with normal use. In the examples shown it is assumed that all the excavated material from the patio can be disposed of in the garden.

		⏱	£	Skill Factor	£
Patio size 4×3 m					
Excavate to remove vegetable soil 150 mm and spread in garden	12 m²	6:00	–	2	
Bed of sand 100 mm thick	12 m²	3:00	6.00	2	
Precast concrete natural colour paving paving slabs, 600×600×50 mm thick	12 m²	8:00	102.00	5	
		17:00	108.00		400.00

This table shows that it would probably take you about 17 man hours to lay the patio with a material cost of £108.00 and a builder would charge £450 for the work. If you wish to lay a different material, the table below sets out the cost of alternative materials and you would need to substitute these prices for the precast concrete slabs above to obtain an overall rate for paving. Further examples are also set out.

	Unit	Material £
Precast concrete natural colour paving slabs, size		
450×450×50 mm thick	1 m²	12.00
600×600×50 mm thick	1 m²	8.50
750×600×50 mm thick	1 m²	8.00
900×600×50 mm thick	1 m²	7.50
Precast concrete coloured paving slabs, size		
450×450×50 mm thick	1 m²	13.00
600×600×50 mm thick	1 m²	9.50
750×600×50 mm thick	1 m²	9.00
900×600×50 mm thick	1 m²	8.50
Reconstructed stone textured paving slabs, size		
300×300×50 mm thick	1 m²	24.00
300×450×50 mm thick	1 m²	18.00
450×450×50 mm thick	1 m²	16.00
Brick paving (£450/1000)		
straight joints laid flat	1 m²	24.00
straight joints laid on edge	1 m²	34.00
herringbone pattern laid flat	1 m²	24.00
herringbone pattern laid on edge	1 m²	34.00
Granite setts		
100×100×75 mm	1 m²	50.00

Here are some examples of different sized patios using some of the materials listed above.

				Skill Factor	
Patio size 4×4 m					
Excavate to remove vegetable soil 150 mm and spread in garden	16 m²	8:00	–	2	
Bed of sand 100 mm thick	16 m²	2:00	8.00	2	

	📏	🕐	💷	Skill Factor	💷
Precast concrete natural colour paving paving slabs, 450×450×50 mm thick	16 m²	12:00	192.00	5	
		22:00	200.00		600.00
Patio size 6×4 m					
Excavate to remove vegetable soil 150 mm and spread in garden	24 m²	12:00	–	2	
Bed of sand 100 mm thick	24 m²	6:00	12.00	2	
Precast concrete natural colour paving paving slabs, 900×600×50 mm thick	24 m²	12:00	288.00	5	
		30:00	300.00		900.00
Patio size 6×6 m					
Excavate to remove vegetable soil 150 mm and spread in garden	36 m²	18:00	–	2	
Bed of sand 100 mm thick	36 m²	9:00	18.00	2	
Precast concrete coloured paving slabs, 450×450×50 mm thick	36 m²	27:00	468.00	5	
		54:00	686.00		1,200.00
Patio size 8×4 m					
Excavate to remove vegetable soil 150 mm and spread in garden	32 m²	16:00	–	2	
Bed of sand 100 mm thick	32 m²	8:00	16.00	2	
Precast concrete coloured paving slabs, 750×600×50 mm thick	32 m²	42:40	288.00	5	
		66:40	304.00		1,350.00
Patio size 6×6 m					
Excavate to remove vegetable soil 150 mm and spread in garden	36 m²	18:00	–	2	
Bed of sand 100 mm thick	36 m²	9:00	18.00	2	

	⏚	£	Skill Factor	
Reconstructed stone textured paving slabs, 450×450×50 mm thick	36 m²	27:00	576.00	5	
		54:00	654.00		1,500.00
Patio size 8×4 m					
Excavate to remove vegetable soil 150 mm and spread in garden	32 m²	16:00	–	2	
Bed of sand 100 mm thick	32 m²	8:00	16.00	2	
Brick paving (£450 per 1000) laid flat	32 m²	42:40	768.00	5	
		66:40	784.00		1,750.00

WALLING

Building brick or stone walling in gardens is either to mark a boundary or for decorative purposes. In these examples it is assumed that the excavation for the foundations will be done by hand and that the concrete is mixed off site and delivered and deposited within 25 metres of its point of placing.

If the surplus excavated material is not deposited on site, see *Disposal of Material* page 102, for the cost of skip hire.

	⏚	£	Skill Factor	
Preparatory work					
Excavate trench size 450×225 mm deep for wall foundation	1 m	0:25	–	2	
Concrete in wall foundation	1 m	0:20	6.50	2	
per metre		0:45	6.50		10.00

The above costs and times should be added to various types of walling set out below to establish the total and costs for each type of wall.

			£	Skill Factor	
Half brick thick wall 112 mm thick in common bricks laid in cement mortar					
£150 per thousand	1 m²	2:00	14.00	6	46.00
£180 per thousand	1 m²	2:00	16.00	6	50.00
£220 per thousand	1 m²	2:00	18.00	6	54.00
One brick thick wall 225 mm thick in common bricks laid in cement mortar					
£150 per thousand	1 m²	2:00	28.00	6	46.00
£180 per thousand	1 m²	2:00	32.00	6	50.00
£220 per thousand	1 m²	2:00	36.00	6	54.00
One brick thick wall 225 mm thick in facing bricks laid in cement mortar					
£250 per thousand	1 m²	2:00	40.00	6	92.00
£300 per thousand	1 m²	2:00	46.00	6	100.00
£350 per thousand	1 m²	2:00	52.00	6	108.00
£400 per thousand	1 m²	2:00	60.00	6	116.00
£450 per thousand	1 m²	2:00	64.00	6	122.00
£500 per thousand	1 m²	2:00	66.00	6	130.00
Random rubble stone laid dry, thickness					
300 mm	1 m²	3:30	50.00	7	90.00
450 mm	1 m²	4:00	70.00	7	120.00
500 mm	1 m²	4:50	80.00	7	130.00
Random rubble stone laid in gauged mortar, thickness					
300 mm	1 m²	4:00	55.00	7	105.00
450 mm	1 m²	4:30	75.00	7	135.00
500 mm	1 m²	5:00	85.00	7	145.00

TIMBER AND DAMP TREATMENT

Timber treatment

It is said that more than half the houses in the UK suffer from insect infestation, the most common being the furniture beetle. Fortunately, they only do minor damage, mainly to unpainted wood surfaces, but the problem can be serious if they infest structural timbers.

If the attack is not too widespread, you can probably deal with it yourself by treating the affected areas (and those not yet affected) by applying an insecticidal wood preservative by brush.

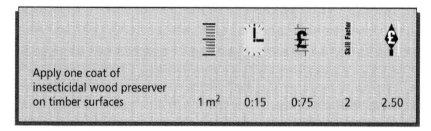

				Skill Factor	
Apply one coat of insecticidal wood preserver on timber surfaces	1 m²	0:15	0.75	2	2.50

If you have decided to use spraying equipment instead of brushing by hand, you can hire the equipment for about £12 for the weekend.

Damp treatment

There are two main types of dampness in houses, penetrating and rising damp. Both types can cause serious long-term damage to a building coupled with unpleasant and unhealthy living conditions before the problem is solved.

Penetrating damp is usually caused by one (or more) of the following conditions:

- cracked or porous brickwork;
- defective rendering;
- damaged flashing;
- leaking gutter or rainwater pipe;
- bridged cavity.

Here are the times and costs for the remedial work to cure these defects.

				Skill Factor	
Rake out joints of brick walls and point up in mortar	1 m²	1:20	0.75	5	12.00
Rake out joints of chimney stacks and point up in mortar	1 m²	1:30	0.75	5	12.00
Cut out single brick from external wall and replace with new brick bedded in cement mortar	1 no	1:00	0.50	5	8.00

		🕐	£	Skill Factor	🔧
Rake out joint in brick wall, re-fix flashing and point up in mortar	1 m	0:45	0.50	5	8.00
Hack off defective rendering and renew with 19 mm thick cement/sand render	1 m²	1:30	3.80	7	14.00
Remove existing rainwater gutters and fittings, fix new length of gutter cast iron, diameter					
76 mm	1 no	1:30	17.00	5	35.00
115 mm	1 no	1:45	20.00	5	40.00
PVC-U, diameter					
75 mm	1 no	1:00	7.00	5	25.00
110 mm	1 no	1:30	9.00	5	35.00
outlet	1 no	1:00	5.00	5	7.00
stop end	1 no	1:00	3.00	5	4.00
Remove existing rainwater pipe and fittings, fix new length of pipe cast iron, diameter					
75 mm	1 no	1:30	35.00	5	55.00
100 mm	1 no	1:45	48.00	5	70.00
PVC-U, diameter					
68 mm	1 no	1:00	12.00	5	25.00
110 mm	1 no	1:30	20.00	5	35.00
Remove six bricks from external leaf of cavity wall, inspect cavity and remove any bridging between leaves, reset bricks and point up	1 no	2:30	1.50	5	30.00

Rising damp

The cause of rising damp is usually a defective damp-proof course (DPC). There are methods of replacing DPCs but they are expensive and slow. The most effective way, particularly for a DIY enthusiast, is to inject waterproofing liquid into the wall.

This is done by hacking off the plaster from the internal face of the wall and removing the skirting board. Then holes are drilled about 100 mm apart through each face of the wall and the nozzles inserted. The liquid is pumped into the wall and saturates the brickwork to make it waterproof.

The nozzles are usually in sets of six and it should be possible to treat an average semi-detached house in a weekend if the preparatory work was done first. The cost of hiring the drilling and pumping equipment is about £60 per day, £110 per week and £70 for a weekend.

				Skill Factor	
Drill 18 mm diameter Holes at 100 mm centres into brickwork, inject waterproofing liquid and make good	1 m	2:30	3.50	5	15.00
Hack off plaster from brickwork and renew	1 m²	2:40	3.50	8	14.00
Take off existing timber skirting board 100 m high and renew	1 m	0:40	2.00	6	5.00

DISPOSAL OF MATERIAL

The items for demolitions, alterations and excavation in this book do not include the cost of removing the debris or excavated material. There are two ways of doing this, either by the use of skips or loading into lorries.

In domestic situations, it is unlikely that lorries would be used so the figures below refer to shifting rubbish by loading it into skips. It is assumed that the skips would be fully laden and the prices include VAT.

Skip size tonnes	Cost £	Cost per cubic metre
3	30.00	10.00
4	45.00	11.25
5	55.00	11.00
6	65.00	10.83
8	85.00	10.63

These costs show that there is not much difference in price per cubic metre for different sized skips so it may be better to order a slightly larger skip than you need just in case the material bulks, more than you anticipated, after excavation. For all practical purposes, one tonne is the equivalent of one cubic metre but it depends to some extent on the nature of the material being dumped.

Skip hire firms used to charge by the day but they seem to take a much more relaxed view these days and are prepared to leave the skip with you for up to a week or more if you ask without any extra charge. The following figures show the cost of moving the rubbish to the skip.

				Skill Factor	
Load into wheelbarrow, wheel 25 metres, and deposit in skip	1 m³	1:20	–	1	9.30

These prices may need adjusting to suit the price levels in your area.
See page xi in the introduction on how to adapt them
for your particular part of the country

PART FOUR

BUILDING YOUR OWN HOME

Why self-build?	107
Finding a plot	108
Finding the money	112
Professional help	113
Preparing a budget	115
Making a success of self-building	118

4

BUILDING YOUR OWN HOME

WHY SELF-BUILD?

Every year over 25 000 people build their own homes and about 250 000 think about it! Contrary to popular belief, most people who achieve their ambition do not have lots of money and do not have any previous knowledge of the building industry.

Self-builders have different motives for making the decision to build their own homes. Some do it to express their own individuality and create something different. Others wish to make their contribution to the conservation movement by using eco-friendly materials. But all self-builders hope to save money by doing it themselves rather than buying a house from a builder or developer.

Although this is often the case, there is danger of underestimating the commitment needed both in time and money to build a house. Anything less than a total dedication to the project will cost you financially and the benefits looked for in the first place will disappear.

It cannot be overstressed that the most successful self-built projects are those that have been planned and budgeted for, in the most detail. Planning the sequence and order of the work can identify problems and difficulties whose impact can be reduced, or even prevented, with a little forethought.

The subject of self-building is set out below under the following headings:

- finding a plot;

- finding the money;

- professional help;

- preparing a budget; and

- making a success of self-building.

There are three main types of self-builders. First, is the hands-on type who wants to do as much of the work as possible. Couples working together usually come into this category and they are prepared to work weekends and holidays for two to three years to fulfil their dreams.

These people gain the most from the self-build experience because of the immense feeling of satisfaction they receive on completing the project. The financial rewards are substantial – a self-builder providing all his own labour can save as much as 40 per cent of the commercial value of the property on completion.

The second type is the opposite of the first. This couple do not have the time, skill or inclination to build their house and are prepared to let someone else do it for them. If they want a traditional bricks-and-mortar house they will probably engage an architect who will draw up the plans, obtain the necessary approvals, appoint a contractor and supervise the building works.

Alternatively, the couple may approach a firm specialising in timber-framed houses and choose one of hundreds of house designs available. These firms offer a range of services from supplying only the frame to building the complete house and handing over the key when it is finished – sometimes called a turnkey contract for obvious reasons.

The most common type of self-builder lies in the third category. This is the type who has not much practical knowledge or skill but is able to manage the construction of a house. He will have the plans drawn up and appoint separate contractors to carry out packages of clearly-defined work.

He will plan, co-ordinate and supervise the work and, although the financial rewards will not be great as doing all the work himself, he will be able to save significant sums by adopting this method of working.

FINDING A PLOT

Most people who have built their own homes have a plot-hunting story to tell. Finding a plot of land is probably the most difficult part of the self-building operation. The first and most important lesson to learn is that you won't find a plot by waiting for it to come and find you!

Anything less than an aggressive whole-hearted approach to the search will almost certainly end in failure. There are various ways to go about finding a plot and some of them are listed below but perseverance is essential however you decide to go about it.

Registering an interest with someone is a start but you must call them regularly to let them know you are still looking. Make a nuisance of yourself in a polite way and you may find that they will get rid of you by putting you to the front of the queue!

Local authorities

Some authorities keep a register of land for self-builders in an attempt to create a balance of different types of properties in their areas. Pay regular visits to your local planning office and see who is applying for planning permission and for what.

Look particularly at applications from people in large houses looking for outline planning permission to build a house in their garden. Have a look at the house and if it looks a little run down and in need of repair, it may be that the occupant is trying to raise some capital by selling off part of the garden.

The best approach then is to drop a note through the door explaining that you have seen the application and wonder whether you could talk about the possibility of buying the land. Leave a telephone number and see what happens. It is better not to approach the vendor face-to-face at this stage – he or she may be nervous and would not welcome such a direct approach.

Local authorities can also be a good starting point for enquiries about the ownership of plots and land that appear to be neglected and derelict.

Estate agents

Selling plots of land is a peripheral activity for most estate agents but plots are bought and sold through their contacts with the general public and developers. You should pay regular visits to agents in your area in person and by telephone.

Make friends with the staff and if anything does come on the market they may contact you first. Remember that you can't embarrass an estate agent by repeatedly asking whether he has any plots for sale! That's what he is there for.

Agents are usually one of the first people to know about future housing developments by developer clients. Sometimes a developer cannot afford to buy a piece of land outright and may be prepared to sell-off a couple of plots to self-builders to enable him to make the deal. In these circumstances, having a self-builder on his register keen to buy a plot, could help the agent secure his own position with the developer.

Developers

The situation described above where the developer cannot afford to buy a piece of land because of its size can occur and an agent may not be involved. Sometimes developers buy plots of land and store them in a land bank for rainy day. If they have cash flow problems (and house development can be a very bumpy ride), they may be prepared to sell off some plots to tide them over.

Write to developers in the area where you would like to live and explain that you are looking for a plot for a four-bedroom house or whatever and state the price range you can afford. You may get a pleasant surprise!

Plot hunters

Because of the number of people looking for plots of land, a new type of agent has appeared. These are firms who specialise in identifying parcels of land for sale and they all claim to provide a fast efficient service. Speed is essential in this context because decent plots are sold very quickly unless they are ridiculously overpriced.

One of the leading firms operating in this field is Plotsearch (0870 8709994) who provide a comprehensive plot finding service including a facility to look at aerial pictures of individual plots on their website (www.homecheck.co.uk). This service costs £20 for six months access to their aerial photographic files, on top of the £39 fee for access to the rest of their register of sites.

The only drawback to using firms like Plotsearch is that other people also have access to the same data so the need to be quick off the mark is essential. You won't be the only person examining their files!

Timber frame companies

Firms who sell timber-framed houses in kit form have a vested interest in having access to plot availability information. If they think you are a genuine prospective customer for their products they will try their best to find a plot to suit your needs. Visit www.buildit.online.co.uk for a list of firms in this category.

It would help if you could identify the type of plot you are looking for. For example, telling them that you wanted a plot for a four-bedroom detached house in South-West Surrey, preferably in a rural setting, would narrow the field of search. But don't be too pedantic in case you specify yourself out of the market!

Existing houses

Many self-builders have achieved their ambition by purchasing an existing property, demolishing it and building a new house in its place. The local planners will take a keen interest in this type of development but if they are consulted all during the process they are usually sympathetic.

After all, it is in the local authorities' interest to have ratepayers in a new house rather than receiving no income from a derelict cottage. The restrictions usually imposed are concerned with the size of the new house compared to the previous one but a good planning consultant would probably be able to achieve what you want by negotiating on your behalf with the planners.

Estates

There are many estates in the UK large enough to be managed by their own estate departments. The owners of these estates range from the landed gentry to business conglomerates but they all have one thing in common. They all own property in the form of cottages and houses and are also determined to run their estates as efficiently as possible.

It is worth writing to estate managers in the area where you want to live and ask to be kept informed about the sale of any properties in the future.

Plot sharing

Sometimes plots come on the market that are too large (and too expensive) for individual purchasers. It is worth checking through websites or advertising in your local papers to find people who would be interested in sharing the cost of buying such a plot so that the plot could be developed jointly.

An informal arrangement could be made with several different people on this basis and this would increase your chances of finding a plot. Alternatively, if you have already chosen a design from a package company and found a plot large enough for two or three houses, they may well be interested in buying the site. This would allow them to make a sale to you and probably to other customers.

The only down side to plot sharing is the danger of entering into a serious business arrangement with people you don't know too well and having them as neighbours afterwards! You should only go down this route with the advice of a solicitor behind you.

Other ways

There are various other methods of looking for a plot and they all involve pursuing sources that may not seem too promising at first. I know one couple who are now on their third self-build project and they found all three plots by putting postcards in their local supermarkets!

The cards were handwritten (to show it was an individual not a firm advertising) and all had the same message.

'HAVE YOU A LARGE GARDEN? WOULD YOU LIKE MORE MONEY? MY WIFE AND I WOULD LIKE TO BUY SOME LAND TO BUILD OUR DREAM HOUSE – PLEASE CALL OUR NUMBER...'

This approach worked three times! He said he was also going to get some leaflets printed to put in the doors of large houses in the area but was advised that it looked too professional. So they decided to use handwritten letters instead but, in the event, they had no need to.

Advertising in local papers should not be overlooked but it can be expensive in the long term. Making your needs known to as many people as possible is a good way of spreading the work, particularly if you offer some form of commission if anyone can find the land for you.

Visit your planning office and look for applications for outline planning over a year old. If the work hasn't gone ahead, there may have been a change in circumstances and a letter to the owner could be worthwhile.

Summary

The options listed above are not meant to be alternatives to each other. If you are serious about building your own house you should be trying all or most of them at the same time. You will almost certainly be disappointed in your search at some time unless you are lucky, but you have to keep looking and looking in order to find the plot that suits you.

Good plots are sold quickly so you must have your financial arrangements in place so that you can move quickly when the right plot comes along.

FINDING THE MONEY

There are various ways of raising money to pay for a plot of land and the building costs, and you should research the options thoroughly before going too far down the self-build road. Banks and building societies are usually willing to help but you must make sure that you are getting the best deal possible for yourself. Different sets of criteria are applied by different lenders so you have to shop around.

Most lenders set a borrowing limit based on a multiplier of 2.5 on a joint income or between 3 and 3.75 for a single earner. They also take different views on overtime and bonus earnings and promotion prospects. Their main aim is to make sure that their money is in safe hands and you can afford to make the repayments without difficulty.

The equity in your own house (if you have one) will be part of the equation when you determine how much you need to borrow plus any savings you may have. Some banks and building societies have departments specialising in self-build mortgages and they are obviously going to be more sympathetic to self-build proposals than general-purpose money lenders.

About three years a Scottish firm, Buildstore, introduced the Accelerator mortgage that approached the self-build mortgage market in a completely different and more helpful way. Previously, advances were made on building work after certain agreed stages had been reached; foundations, first floor level, eaves level and the like.

Unfortunately, by the time the payments were made, the men carrying out work and the merchants supplying the materials were unhappy about what they saw as late payments. The danger of losing key men because of this delay in payment was real, but the Accelerator mortgage removed that worry.

This scheme makes the payments in advance of the work on the agreed stages being reached so that the job can proceed smoothly without the risk of any hold ups caused by late payments. This mortgage system will advance 95 per cent of the land cost so that the purchaser is in a strong position when a plot becomes available.

The borrowers then have the choice of staying in their own homes until the work is complete, moving into temporary rented accommodation or moving on site and living in a caravan in order to supervise the work and be on hand to receive material deliveries.

This type of mortgage is becoming more popular because it seems to be driven by what the borrower wants rather than having to put up with arrangements that are more suited to the needs of the lender.

Here are the names and contact points for firms offering the Accelerator mortgage including the originator Buildstore. This information is reprinted with the permission of *Build It* magazine. For further information on any of these lenders call 0870 872 0908.

Lender	Advance on land (%)	Advance on final value of property (%)	Area covered
Buildstore	95	95	National
Britannia BS	95	95	National
Lloyds TSB	95	95	Scotland
Skipton BS	95	95	Eng/Wales/Scot
The Mortgage Business	95	85	National
Verso	95	85	National

PROFESSIONAL HELP

If you are buying a plot of land without outline planning permission, you will need the services of a planning consultant, particularly if there are any potential problems involved. Most plots, however, are sold with planning permission already obtained so the first professional you will need to employ will be someone to prepare the plans and submit them for detailed planning permission.

If you intend using a firm specialising in package deals, this will usually be part of their service. All you have to do is choose the design and they will provide the drawings and take them through the planning procedures on your behalf. But some people want their new home to be unique and the only way to achieve that is by transferring your ideas and needs on to paper. Very few people have the skill and resources to do this themselves so it is necessary to look for professional help.

The architectural profession has changed dramatically in the last 20 years. Previously, architects worked to a scale of fees and were in serious trouble with their professional body if they deviated from it. Those days have gone and, although the fee scales still exist, they are not mandatory any more. This means that you can shop around, not only for the architect whose style of design you admire, but also for the cost of his services.

In your initial talks with an architect, ask how much his fees will be. If he quotes you hourly rates be careful. Quoting £50 per hour for his time and £20 for a technician may be acceptable to you but without an assessment of how many hours are involved, these rates are meaningless.

And if he is going to apply some notional hours to the rates, why not go all the way and offer a lump sum figure? Paying by the hour can work but it is like handing someone a blank cheque and trusting them to keep their costs down to a minimum. You wouldn't do it for a firm digging your drains so why do it for an architect?

Often the architect's fee is expressed as a percentage of the cost of the building work and this could be as much as 12 per cent for a full service including supervising the work, or about 5 per cent for pre-contract services. If you are going to shop around for an architect, ask for a quote based on a percentage because it is easier to compare percentages than hourly rates where the number of hours is not known. If you decide to pay on a percentage basis, make it absolutely clear which parts of the work are subject to the agreed percentage and which are excluded

Don't overlook the possibility of employing an architectural technician to prepare your drawings. You will probably see them advertising their services under the heading of 'Plans Drawn' in your local paper. See what other work they have carried out and talk to their previous clients. Their hourly rate would probably be between £15 and £20 an hour and ask them for a likely overall cost.

At some stage of the design process, you will need the services of a structural engineer to do the calculations for beams, lintels and the like. Usually, the person preparing the drawings will employ him and pass the charge on to you or you may be asked to pay the fees yourself direct. It will not be a major expense but it needs to be discussed before work on the drawings commences. You want any nasty surprises early in the project! It would be unusual to employ a quantity surveyor in the construction of a one-off house. He would probably charge about 3 per cent of the cost of the building work for preparing a full bill of quantities and 5 per cent for a full service including assessing monthly interim payments and settling the final account. It would be a little extravagant to use a quantity surveyor on a small project but there is one service he can provide that could be extremely valuable to a self-builder.

For a fee of between 1 and 1.5 per cent, a QS would prepare an approximate estimate of the cost of the work for the job from the completed drawings and also negotiate with the contractor if necessary. This estimate would be the cornerstone of the budget and could be an invaluable tool in negotiations with contractors and sub-contractors. Further, it would show mortgage lenders that proper attention was being paid to the financial side of the project.

PREPARING A BUDGET

It is a paradox that the time you need to have the most accurate cost information is the time when it is least likely to be available to you! This is when you are at the first stages of planning the project. Different people have different motives for building their own houses. Some are profit driven (not always successful), some because they need more space and can't afford to buy a larger house and some because they just want to express themselves!

Whatever the motive, it is essential that a budget is prepared and maintained from the beginning to the end of the project. If something is about to go wrong, you want to know about it as soon as possible so that you can take corrective action.

For example if, despite all your efforts of shopping round, the quotes you receive for roof tiling are £5000 higher than you allowed for, you will have to look for savings elsewhere. Perhaps by postponing some of the landscaping work or changing the specification of the driveway. Whatever you decide to do you will be able to keep financial control adjusting your budget.

A budget starts off as a rough-and-ready assessment of your overall plan and gradually becomes more accurate as cost information is fed into it. There are two ways in which a budget for a self-build project starts life. The first is when someone assesses how much money and equity they have and how much they can borrow and wonder if they can build a new house for that sum. The second approach is when a larger house is needed, usually for family reasons, and the calculation is then reversed – how much would it cost to build a house sized say, 2000 square feet?

In both cases it is necessary to get a feel for the size of houses. Start by measuring up the total area of your own house. The easiest way is to do it from the outside and make one foot deduction for each external wall and multiply the length by the width to produce the approximate ground floor area and multiply by two for the first floor. This will give you the approximate area and you can then judge your future needs by applying the area of existing house as yardstick.

You can also look at literature from show houses in your area and compare the stated area of the houses with the accommodation provided. So a budget generally starts off with a lump sum figure representing your estimated assets and how much you can borrow.

Let us assume this figure comes to £120000. From this you must deduct an amount to buy a plot. There is a theory that the plot value should not exceed 30 per cent of the building cost but each plot has its own good and bad characteristics so you should judge each one on its own merits. In this case, it would be reasonable to be looking out for a plot between £30000 and £40000.

Say the plot costs £30000 so that leaves £90000 for the construction costs but you should put about £10000 to one side for fees and contingencies. The question then is – can you build a house for £80000? The answer depends on how big the house is. Here is a table showing the range of areas available for £80000.

Cost per square foot	Area square feet
£50	1 600
£55	1 454
£60	1 333
£65	1 230
£70	1 142

There are many examples of people building their own houses for £50 per square foot and even less. Working to a reasonable standard of finish, it should be possible to manage the building of your house for between £50 and £60 per square foot but you must expect to pay more if you employ a builder to carry out all the work. Another approach is to take the £80000 to a package company and see what they can offer. You will probably be surprised by the range and quality of the designs available.

The square foot breakdown for other sums of money are shown below.

Amount available £	Cost per square foot £	Area square feet
90 000	50	1 800
90 000	55	1 636
90 000	60	1 500

Amount available £	Cost per square foot £	Area square feet
90 000	65	1 384
90 000	70	1 285
100 000	50	2 000
100 000	55	1 818
100 000	60	1 666
100 000	65	1 538
100 000	70	1 428
120 000	50	2 400
120 000	55	2 200
120 000	60	2 000
120 000	65	1 846
120 000	70	1 714
150 000	50	3 000
150 000	55	2 750
150 000	60	2 500
150 000	65	2 307
150 000	70	2 142

If you are going to build or manage the building yourself, the next stage in budgeting is to draw up a list of packages of work and place values against them. Here is a list of elements with notional percentages set against them.

Element	%
Site clearance and foundations	8
External walls	15
Roof	14
Floors	4
Internal walls and partitions	6
Windows and doors	12
Finishes	10
Fittings	5
Plumbing	5
Heating	4
Electrical work	4
Drains and external works	13
	100

When this elemental cost breakdown or list of work packages is prepared the budget becomes an invaluable tool in providing financial control of the project. The value of each element or package should be updated as new and better information is available so that a theoretical final account figure can be produced at any one time. Apart from bringing piece of mind, it means that potential underspending and overspending can be identified at an early stage so that corrective action can be taken.

MAKING A SUCCESS OF SELF-BUILDING

It is rare to hear of an unsuccessful self-build project and the reasons for this are not hard to find. Nearly all self-builders have to go through a long, character-forming process in finding land and the dreamers and the half-hearted soon lose interest.

The people who eventually find the plot they have been looking for are generally determined, single-minded individuals who are not going to let a simple task of building a house worry them! In most cases, the task of getting down to the planning and building or managing the construction of their dream home, is a relief after the frustrations and disappointments of looking for land. But here is a danger, that in a rush to do something practical, the need for careful planning is neglected causing mistakes that can influence events when the building work starts.

There is a common thread running through most self-build projects and the following advice has been taken from those who have successfully built their own houses.

- Listen to the experts but don't take every piece of advice literally. They are generally talking in broad terms based upon their previous experience and this may not necessarily apply to your needs.

- The longer you spend planning the project, the better chance there is of things going well.

- Either stay in your own home until your new one is ready or live in a caravan on site. Don't sell your house and move into rented accommodation – it will cost you!

- Give a lot of thought on how you will accept the delivery of materials. It can be a major problem if you are not living on site.

- Keep records! Note the significant points in any conversation you have before and during the progress of the works. Keep a diary and take photographs from about six different positions every week.

- Spend as much time as it takes to brief the person designing your house. Collect pictures from magazines showing the overall effect you would like to achieve. Think what your needs will be in five years' time and build those into the drawings.

- Prepare a budget and make it work for you and update it regularly. Look for trends of under- and overspending and act accordingly.

- Obtain at least three tenders for each package of work. Be prepared to negotiate on costs but build timing and programming into the acceptance. Avoid contractors who want paying in cash unless for very small amounts of work.

- Agree stage payments and pay on time. Build in a small retention amount of say, 5 per cent, for defects but release money after three months.

- Ask contractors already working on site if they can recommend other trades.

- Hold regular meetings on site to discuss progress and quality of work.

- Make sure you have adequate insurance for public and employers liability, accident, theft, injury, malicious damage and any other special needs your site may have.

- Consider buying a second-hand temporary shed-like building for storage that can be sold afterwards.

- Avoid making changes to the design once the work starts. Variations will cost you both time and money.

- Confirm in writing any important changes to the scope or timing of the work that has been agreed with contractors as soon as possible.

- Don't expect that everything will go smoothly. Problems arise on every building site and your job is to identify and solve them.

PART FIVE

TOTAL PROJECT COSTS

Traditional house extensions   123

PVC-U conservatories   125

Loft conversions   125

Swimming pools   126

5

TOTAL PROJECT COSTS

TRADITIONAL HOUSE EXTENSIONS

Although PVC-U extensions have increased their share of the market in recent years, the traditional bricks and mortar extension is still the most popular. This is partly due to the fact that it is traditional but also because two storey extensions can only be constructed in this way.

The following table shows an elemental breakdown for a typical extension expressed in percentage terms.

Element	%
Preliminaries	15
Substructure to DPC level	12
External walls	23
Flat roof	12
Windows and external doors	14
Wall finishes	6
Floor finishes	3
Ceiling finishes	2
Electrical work	5
Heating work	4
Alterations	4
Total	100

The tables below show the costs per square foot for a variety of extension types and sizes. It can be seen that the larger the extension, the lower the cost per square foot.

External size of extension in feet	Internal area of extension in square feet	Cost per square foot £
One storey extension with flat roof		
7 × 10	48	140
7 × 13	66	105
7 × 16	84	100
10 × 10	72	105
10 × 13	99	85
10 × 16	126	75
10 × 20	162	70
13 × 13	132	80
13 × 13	168	80
13 × 20	216	65
One storey extension with pitched roof		
7 × 10	48	155
7 × 13	66	115
7 × 16	84	110
10 × 10	72	120
10 × 13	99	100
10 × 16	126	85
10 × 20	162	80
13 × 13	132	95
13 × 13	168	85
13 × 20	216	75
Two storey extension with flat roof		
7 × 10	48	115
7 × 13	66	95
7 × 16	84	80
10 × 10	72	90
10 × 13	99	70
10 × 16	126	60
10 × 20	162	55
13 × 13	132	65
13 × 13	168	60
13 × 20	216	50

External size of extension in feet	Internal area of extension in square feet	Cost per square foot £
Two storey extension with pitched roof		
7 × 10	48	120
7 × 13	66	90
7 × 16	84	85
10 × 10	72	90
10 × 13	99	75
10 × 16	126	65
10 × 20	162	60
13 × 13	132	70
13 × 13	168	60
13 × 20	216	55

PVC-U CONSERVATORIES

The PVC-U double glazing market is fiercely competitive and it is difficult to give accurate figures for typical conservatories. Most firms operating in this field work on the principle of '... we'll beat any other similar quote so come back when you have other prices ...'

This approach goes against normal tendering procedures but it should be possible to have a medium-sized conservatory built for about £40 per square foot including foundations but excluding any external works.

LOFT CONVERSIONS

There are three different sizes of loft conversions stated in this section and the following table shows an elemental breakdown for a typical conversion expressed in percentage terms.

Element	%
Preliminaries	16
Preparation	2
Dormer window	38
Roof window	7
Stairs	11

Element	%
Flooring	6
Internal partitions and doors	6
Ceilings and soffits	5
Wall finishes	2
Electrical work	3
Heating work	4
	100

The first problem to overcome when considering building a loft conversion is the question of access. If a staircase cannot be accommodated easily in the existing first floor room layout, it is not worth proceeding. There are stringent building regulations to be observed and leaflets should be obtained from your local planning office. The table below show the costs per square foot for three loft conversion projects.

Overall size of conversion in feet	Internal area of conversion (including dormers) in square feet	Cost per square foot £
15×15	260	45
15×18	310	40
15×21	375	37

SWIMMING POOLS

The cost of swimming pools varies considerably in accordance with the type and size required. The cheapest type is the DIY pool kit range costing about £2000 for a pool 10×20 feet and comprises a series of panels bolted together.

The next type is made of moulded fibreglass that arrives on site in one piece to be placed in a prepared hole in the ground. These range from £4500 (10×8 feet) to £7500 (30×15 feet). Above-ground pools are not too costly at £2000 to £2500 for pools 12 feet diameter.

The dearest are concrete based pools enclosed in a building and a pool 18×10 feet could cost about £40000 supplied and erected.

PART SIX

TOOL AND EQUIPMENT HIRE

6

TOOL AND EQUIPMENT HIRE

The following rates are based on average hire charges made by hire firms in the UK and are intended as a guide. You should check your local dealer for the rates that apply in your area. The prices include VAT.

	Day £	Weekend £	Week £
Concrete and cutting equipment			
Concrete mixers			
petrol, with stand	14.00	18.00	30.00
electric, with stand	13.00	17.00	26.00
Vibrating pokers			
petrol	47.00	60.00	94.00
electric	35.00	44.00	76.00
air poker, 50 mm	30.00	38.00	60.00
air poker, 75 mm	33.00	40.00	60.00
Disc cutters			
electric, 300 mm	24.00	30.00	48.00
two stroke, 300 mm	26.00	32.00	52.00
two stroke, 350 mm	30.00	37.00	65.00
electric wall chasers	40.00	50.00	70.00
Block and slab splitters			
clay	56.00	72.00	112.00
block	26.00	32.00	52.00
slab	42.00	52.00	82.00

	Day £	Weekend £	Week £
Saws			
electric, 150 mm	40.00	50.00	76.00
electric, 300 mm	44.00	55.00	82.00
Access and site equipment			
Ladders			
double ladder, alloy			
4 m	14.00	17.00	28.00
6 m	16.00	20.00	33.00
9 m	18.00	22.00	38.00
triple ladder, alloy			
9 m	16.00	20.00	33.00
roof ladder			
5 m	16.00	20.00	33.00
rope operated			
11 m	38.00	48.00	75.00
13 m	42.00	52.00	86.00
16 m	54.00	68.00	103.00
Props			
shoring props			
type 0	0.00	0.00	4.00
type 1	0.00	0.00	4.00
type 2	0.00	0.00	4.00
type 3	0.00	0.00	4.00
type 4	0.00	0.00	4.00
Trestles and staging			
staging			
2.4 m	14.00	17.00	28.00
3.6 m	22.00	26.00	42.00
4.8 m	24.00	30.00	47.00
painters' trestle			
1.8 m	9.00	12.00	21.00
Alloy towers			
single width, height			
2.30 m	35.00	44.00	70.00
3.20 m	42.00	52.00	85.00
3.73 m	44.00	55.00	90.00
4.20 m	50.00	62.00	98.00
5.20 m	66.00	82.00	132.00
6.52 m	70.00	88.00	120.00
7.45 m	68.00	34.00	142.00
7.91 m	87.00	109.00	170.00

	Day £	Weekend £	Week £
8.20 m	94.00	118.00	188.00
9.30 m	101.00	126.00	202.00
10.20 m	120.00	150.00	235.00
full width, height			
2.34 m	42.00	52.00	85.00
3.27 m	47.00	59.00	94.00
3.73 m	50.00	62.00	98.00
4.66 m	58.00	72.00	98.00
5.59 m	64.00	280.00	126.00
6.52 m	68.00	85.00	136.00
7.45 m	75.00	96.00	150.00
7.91 m	78.00	98.00	155.00
8.84 m	87.00	109.00	170.00
10.23 m	96.00	120.00	192.00
11.16 m	106.00	132.00	212.00
12.55 m	122.00	162.00	244.00
13.94 m	136.00	170.00	272.00
14.87 m	144.00	180.00	286.00

Breaking and demolition

Breakers
  hydraulic

diesel	68.00	34.00	142.00
petrol	87.00	109.00	170.00
electric, medium duty	24.00	30.00	47.00
air breaker, medium	35.00	47.00	66.00
air breaker, heavy	38.00	57.00	70.00

Power tools

Drills

cordless drill	18.00	22.00	36.00
cordless impact	21.00	25.00	42.00
two speed impact	9.00	11.00	19.00
rotary drill, 16 mm	26.00	32.00	52.00
rotary drill, 20 mm	28.00	35.00	56.00
combi hammer			
light duty	16.00	20.00	32.00
medium duty	20.00	25.00	47.00
heavy duty	26.00	32.00	52.00

	Day £	Weekend £	Week £
Grinders			
angle grinder			
100 mm	14.00	17.00	28.00
125 mm	14.00	17.00	28.00
230 mm	15.00	18.00	30.00
300 mm	28.00	35.00	56.00
Saws			
reciprocating saw			
standard	20.00	10.00	40.00
heavy duty	22.00	12.00	44.00
circular saw			
150 mm	16.00	20.00	32.00
230 mm	19.00	24.00	38.00
Door trimmer	34.00	42.00	68.00
Woodworking tools			
Plane, 3.25 in	18.00	22.00	35.00
Router	18.00	22.00	35.00
Worktop jig	14.00	17.00	28.00
Fixing equipment			
Cordless nailing gun	28.00	35.00	56.00
Cartridge hammer	24.00	30.00	47.00
Electric screwdriver	14.00	17.00	28.00
Sanders			
Floor	38.00	57.00	70.00
Floor edger	28.00	35.00	56.00
Orbital	18.00	22.00	35.00
Pumps			
Submersible			
25 mm	16.00	20.00	33.00
50 mm	35.00	44.00	70.00
Diaphragm pump, 50 mm	58.00	72.00	98.00

PART SEVEN

GENERAL CONSTRUCTION DATA

The metric system   135

Conversion equivalents (imperial/metric)   136

Conversion equivalents (metric/imperial)   136

Temperature equivalents   137

Areas and volumes   138

General building information   138

7

GENERAL CONSTRUCTION DATA

THE METRIC SYSTEM

Linear	1 centimetre (cm)	= 10 millimetres (mm)
	1 decimetre (dm)	= 10 centimetres (cm)
	1 metre (m)	= 10 decimetres (dm)
	1 kilometre (km)	= 1000 metres (m)
Area	100 sq millimetres	= 1 sq centimetre
	100 sq centimetres	= 1 sq decimetre
	100 sq decimetres	= 1 sq metre
	1000 sq metres	= 1 hectare
Capacity	1 millilitre (ml)	= 1 cubic centimetre (cm^3)
	1 centilitre (cl)	= 10 millilitres (ml)
	1 decilitre (dl)	= 10 centilitres (cl)
	1 litre (l)	= 10 decilitres (dl)
Weight	1 centigram (cg)	= 10 milligrams (mg)
	1 decigram (dg)	= 10 centigrams (mcg)
	1 gram (g)	= 10 decigrams (dg)
	1 decagram (dag)	= 10 grams (g)
	1 hectogram (hg)	= 10 decagrams (dag)

CONVERSION EQUIVALENTS (IMPERIAL/METRIC)

Length	1 inch	= 25.4 mm
	1 foot	= 304.8 mm
	1 yard	= 914.4 mm
	1 yard	= 0.9144 m
	1 mile	= 1609.34 m
Area	1 sq inch	= 645.16 sq mm
	1 sq foot	= 0.092903 sq m
	1 sq yard	= 0.8361 sq m
	1 acre	= 4840 sq yards
	1 acre	= 2.471 hectares
Liquid	1 lb water	= 0.454 litres
	1 pint	= 0.568 litres
	1 gallon	= 4.546 litres
Horsepower	1 hp	= 746 watts
	1 hp	= 0.746 kW
	1 hp	= 33000 ft.lb/min
Weight	1 lb	= 0.4536 kg
	1 cwt	= 50.8 kg
	1 ton	= 1016.1 kg

CONVERSION EQUIVALENTS (METRIC/IMPERIAL)

Length	1 mm	= 0.03937 inches
	1 centimetre	= 0.3937 inches
	1 metre	= 1.094 yards
	1 metre	= 3.282 ft
	1 kilometre	= 0.621373 miles
Area	1 sq millimetre	= 0.00155 sq in
	1 sq metre	= 10.764 sq ft
	1 sq metre	= 1.196 sq yards
	1 acre	= 4046.86 sq m
	1 hectare	= 0.404686 acres
Liquid	1 litre	= 2.202 lbs
	1 litre	= 1.76 pints
	1 litre	= 0.22 gallons
Horsepower	1 watt	= 0.00134 hp
	1 kw	= 134 hp
	1 hp	= 0759 kg m/s
Weight	1 kg	= 2.205 lbs
	1 kg	= 0.01968 cwt
	1 kg	= 0.000984 ton

TEMPERATURE EQUIVALENTS

To convert Fahrenheit to Celsius deduct 32 and multiply by 5/9.
To convert Celsius to Fahrenheit multiply by 9/5 and add 32.

Fahrenheit	Celsius
230	110.0
220	104.4
210	98.9
200	93.3
190	87.8
180	82.2
170	76.7
160	71.1
150	65.6
140	60.0
130	54.4
120	48.9
110	43.3
100	37.8
90	32.2
80	26.7
70	21.1
60	15.6
50	10.0
40	4.4
30	−1.1
20	−6.7
10	−12.2
0	−17.8

AREAS AND VOLUMES

Figure	Area	Perimeter
Rectangle	Length × breadth	Sum of sides
Triangle	Base × half of perpendicular height	Sum of sides
Quadrilateral	Sum of areas of contained triangles	Sum of sides
Trapezoidal	Sum of areas of contained triangles	Sum of sides
Trapezium	Half of sum of parallel sides × perpendicular height	Sum of sides
Parallelogram	Base × perpendicular height	Sum of sides
Regular polygon	Half sum of sides × half internal diameter	Sum of sides
Circle	$pi \times radius^2$ or $pi \times 2 \times radius$	$pi \times diameter$
Cylinder	$pi \times 2 \times radius^2 \times length$ (curved surface only)	$pi \times radius^2 \times length$
Sphere	$pi \times diameter^2$	$Diameter^3 \times 0.5236$

GENERAL BUILDING INFORMATION

Brickwork and blockwork

Bricks per m^2 (brick size $215 \times 103.5 \times 65\,mm$)	
Half brick wall	
Stretcher bond	59
English bond	89
English garden wall bond	74
Flemish bond	79
One brick wall	
English bond	118
Flemish bond	118
One and a half brick wall	
English bond	178
Flemish bond	178
Two brick wall	
English bond	238
Flemish bond	238
Metric modular bricks	
$200 \times 100 \times 75\,mm$	
90 mm thick	133

190 mm thick	200
200 × 100 × 100 mm	
90 mm thick	50
190 mm thick	100
290 mm thick	150
300 × 100 × 75 mm	
90 mm thick	44
300 × 100 × 100 mm	
90 mm thick	50

Blocks per m² (block size 414 × 215 mm)

60 mm thick	9.9
75 mm thick	9.9
100 mm thick	9.9
140 mm thick	9.9
190 mm thick	9.9
215 mm thick	9.9

Mortar per m²	Wirecut m³	1 Frog m³	2 Frogs m³
Brick size 215 × 103.5 × 65 mm			
Half brick wall	0.017	0.024	0.031
One brick wall	0.045	0.059	0.073
One and a half brick wall	0.072	0.093	0.114
Two brick wall	0.101	0.128	0.155

	Solid m³	Perforated m³
Brick size 200 × 100 × 75 mm		
90 mm thick	0.016	0.019
190 mm thick	0.042	0.048
290 mm thick	0.068	0.078
Brick size 200 × 100 × 100 mm		
90 mm thick	0.013	0.016
190 mm thick	0.036	0.041
290 mm thick	0.059	0.067

	Solid m³	Perforated m³
Brick size 200 × 100 × 100 mm		
90 mm thick	0.015	0.018

	Solid m³
Block size 440 × 215 mm	
60 mm thick	0.004
75 mm thick	0.005
100 mm thick	0.006
140 mm thick	0.007
190 mm thick	0.008
215 mm thick	0.009

Masonry

Mortar per m² of random rubble walling	m³
300 mm thick wall	0.120
450 mm thick wall	0.160
550 mm thick wall	0.120

Carpentry and joinery

Length of boarding required	m/m²
Board width, 75 mm	13.33
Board width, 100 mm	10.00
Board width, 125 mm	8.00
Board width, 150 mm	6.67
Board width, 175 mm	5.71
Board width, 200 mm	5.00

Roofing

	Lap mm	Gauge mm	Nr/m² m/m²	Battens
Clay/concrete tiles				
267 × 165 mm	65	100	60.00	10.00
	65	98	64.00	10.50
	65	90	68.00	11.30

	Lap mm	Gauge mm	Nr/m^2 m/m^2	Battens
387 × 230 mm	75	300	16.00	3.20
	100	280	17.40	3.50
420 × 330 mm	75	340	10.00	2.90
	100	320	10.74	3.10
Fibre slates				
500 × 250 mm	90	205	19.50	4.85
	80	210	19.10	4.76
	70	215	18.60	4.65
600 × 300 mm	105	250	13.60	4.04
	100	250	13.40	4.00
	90	255	13.10	3.92
	80	260	12.90	3.85
	70	263	12.70	3.77
400 × 200 mm	70	165	30.00	6.06
	75	162	30.90	6.17
	90	155	32.30	6.45
500 × 250 mm	70	215	18.60	4.65
	75	212	18.90	4.72
	90	205	19.50	4.88
	100	200	20.00	5.00
	110	195	20.50	5.13
600 × 300 mm	100	250	13.40	4.00
	110	245	13.60	4.08
Natural slates				
405 × 205 mm	75	165	29.59	8.70
405 × 255 mm	75	165	23.75	6.06
405 × 305 mm	75	165	19.00	5.00
460 × 230 mm	75	195	23.00	6.00
460 × 255 mm	75	195	20.37	5.20
460 × 305 mm	75	195	17.00	5.00
510 × 255 mm	75	220	18.02	4.60
510 × 305 mm	75	220	15.00	4.00
560 × 280 mm	75	240	14.81	4.12
560 × 280 mm	75	240	14.00	4.00
610 × 305 mm	75	265	12.27	3.74
Reconstructed stone slates				
380 × 250 mm	75	150	16.00	3.20

Plastering and tiling

Plaster coverage	m² per 1000 kg
Carlite browning, 11 mm thick	135–155
Carlite tough coat, 11 mm thick	135–150
Carlite bonding, 11 mm thick	100–115
Thistle hardwall, 11 mm thick	115–130
Thistle dri-coat, 11 mm thick	135–135
Thistle renovating, 11 mm thick	115–125
Tile coverage	**nr per m²**
152×152 mm	43.27
200×200 mm	25.00

Plumbing and heating

Roof drainage	Area m²	Pipe mm	Gutter mm
One end outlet	15	50	75
	38	68	100
	100	110	150
Centre outlet	30	50	75
	75	68	100
	200	110	150

Painting and wallpapering

Average coverage of paints m² per litre	Timber	Plastered surfaces	Brickwork
Primer	10–12	9–11	5–7
Undercoat	10–12	11–14	6–8
Gloss	11–14	11–14	6–8
Emulsion	10–12	12–15	6–10

Wallpaper coverage per roll	Rolls nr	Wall height m	Room perimeter m
	4	2.50	8
	5	2.50	9
	5	2.50	10
	6	2.50	11
	6	2.50	12
	7	2.50	13
	7	2.50	14
	8	2.50	15
	8	2.50	16
	8	2.50	17
	9	2.50	18
	10	2.50	19
	10	2.50	20
	10	2.50	21
	11	2.50	22
	11	2.50	23
	12	2.50	24
	13	2.50	25
	13	2.50	26
	14	2.50	27
	5	2.80	8
	5	2.80	9
	5	2.80	10
	7	2.80	11
	7	2.80	12
	7	2.80	13
	8	2.80	14
	8	2.80	15
	9	2.80	16
	10	2.80	17
	10	2.80	18
	11	2.80	19
	11	2.80	20
	12	2.80	21
	13	2.80	22
	13	2.80	23
	14	2.80	24
	14	2.80	25
	15	2.80	26
	15	2.80	27

External works

Blocks/slabs per m²	nr/m²		
200 × 100 mm	50.00		
450 × 450 mm	4.93		
600 × 450 mm	3.70		
600 × 600 mm	2.79		
600 × 750 mm	2.22		
600 × 900 mm	1.85		

Drainage trench widths	Under 1.5 m deep mm	Over 1.5 m deep mm	
Pipe diameter 100 mm	450	600	
Pipe diameter 150 mm	500	650	
Pipe diameter 225 mm	600	750	
Pipe diameter 300 mm	650	800	

Volumes of filling for pipe beds (m³ per m)	50 mm thick	100 mm thick	150 mm thick
Pipe diameter 100 mm	0.023	0.045	0.068
Pipe diameter 150 mm	0.026	0.053	0.079
Pipe diameter 225 mm	0.030	0.060	0.090
Pipe diameter 300 mm	0.038	0.075	0.113

Volumes of filling for pipe bed and haunching	m³ per m
Pipe diameter 100 mm	0.117
Pipe diameter 150 mm	0.152
Pipe diameter 225 mm	0.195
Pipe diameter 300 mm	0.279

Volumes of filling for pipe bed and surround	m³ per m
Pipe diameter 100 mm	0.185
Pipe diameter 150 mm	0.231
Pipe diameter 225 mm	0.285
Pipe diameter 300 mm	0.391

GLOSSARY

The following glossary of terms is intended to help anyone not familiar with the construction industry to understand the meaning of the words most commonly used.

Aggregate	Small stones or gravel forming part of a concrete mix.
Anaglypta	A heavy embossed wallpaper.
Architrave	A piece of timber covering the joint between a door or window frame and the plasterwork.
Angle bead	A galvanised steel right angle fixed at the vertical external corners of walls to strengthen and protect the plasterwork.
Apron flashing	A flashing placed at the cill of dormer windows or the lower sides of chimney stacks.
Auger	A corkscrew-shaped tool for drilling holes.
Backfilling	Excavated material that is compacted around the sides of foundation walls.
Back putty	A narrow strip of putty between the inside face of a pane of glass and the rebate.
Barge board	A sloping timber board fixed to the gable end of a roof.
Bending spring	A helical coiled spring inserted in copper pipes to protect the pipe walls when bending.
Blinding	The practice of treating the top surface of hardcore or broken bricks with sand to produce a smooth surface to receive concrete.
Bond	The arrangement of bricks or blocks to present an attractive appearance and provide structural strength by staggering the vertical joints.
Bradawl	Hand tool used for making starter holes in wood to receive screws.

Browning coat	A mixture of gypsum plaster, sand and water that acts as backing coat for plasterwork.
Butterfly wall tie	A wall between two leaves of a cavity wall made of galvanised wire formed into a double triangular shape.
Cavity wall	An external wall consisting of two leaves of brickwork of blockwork connected by wall ties.
Close boarded fence	Vertical softwood boards fixed to horizontal softwood rails.
Common brick	A brick usually of poor appearance normally used where it will not be seen, e.g. in foundations or in the inner leaf of a cavity wall.
Coping	The top course of bricks or a concrete slab on top of a wall.
Counter-sinking	A depression made in a timber surface so that the head of a screw can be driven flush with the surface.
De-humidifier	An air-conditioning unit that cools the air to reduce its humidity.
Dowel	A short cylindrical piece of wood or metal sunk into two adjacent members to strengthen the connection.
Eaves tile	A short tile nailed at the eaves of a roof as an extra course.
Efflorescence	An unsightly deposit of crystallised salts on brickwork.
Engineering brick	A dense brick used where strength and durability are required.
Expansion pipe	An overflow pipe from a hot water tank or cylinder usually discharging over a cold water tank.
Fascia	A vertical timber board fixed to the end of rafters.
Flashing	A strip of metal or bitumen felt fixed between roof tiles and brickwork to prevent the ingress of water.
Flaunching	Cement mortar placed around chimney pots to deflect rainwater on to the roof.
Hardcore	Broken bricks or stones laid to receive a concrete bed.
Heave	A swelling in clay caused by excess rainfall.
Jamb	The internal vertical face of an opening or the side member of a door or window.
Lost headed nail	A nail whose head is only slightly larger than its tail that can be driven below the surface.
Luminaire	A light fitting.
Mist coat	A thin coat of emulsion paint applied to the plaster as a first coat.

Mortice	A recess formed in timber to receive a tongue or a tenon from another member.
Newel post	A vertical post in a staircase to support the handrail.
Nogging	A short horizontal timber member fixed between vertical studs.
Overhand	Brickwork to external walls laid from staging inside the building without the use of external scaffolding.
Pargetting	The rendering to the inside of a chimney flue.
Pin kerb	A small pre-cast concrete kerb usually 150 × 50 mm and laid as edging to paths.
Purlin	A horizontal roof member supporting the rafters.
Rendering	The application of mortar to walls.
Reveal	The vertical face of a window or door opening.
Screed	A layer of cement mortar laid on top of concrete slab to receive a floor or roof finish.
Skim coat	The final coat of plaster usually only 3 mm thick.
Spur	A branch from a ring main for a new electric socket outlet.
Studding	A partition constructed of timber.
Tanking	The application of a waterproof membrane to the floor and walls of a basement.
Uncoursed	The arrangement of irregular shaped stones in a stone wall without a continuous horizontal bed.
Underpinning	The technique of replacing a load-bearing wall below ground level by supporting the wall in short lengths and building a new wall.
Verge	The edge of a sloping roof at the gables.
Wall plate	A horizontal timber member set on top of a wall to receive rafters or joists.

USEFUL ADDRESSES

MATERIAL SUPPLIERS

Jeld-Wen
Watch House Lane
Doncaster
South Yorkshire DN5 9LR
(01302 394000)

Woodfit Ltd
Kem Mill Whittle-le-Woods
Chorley PR6 7EA
(01257 266421)

TRADE ASSOCIATIONS AND PROFESSIONAL BODIES

Brick Development Association
Woodside House
Winkfield
Windsor SL4 2DP
Tel: 01344 885651

Builders Merchants Federation
15 Soho Square
London W1V 5FB
(0207–439 1753)

Electric Contractors Association
34 Palace Court
Bayswater
London W2
(0171–229 1266)

Federation of Building Sub-Contractors
82 New Cavendish Street
London W1M 8AD
(0171–580 5588)

Federation of Master Builders
14 Great James Street
London WC1N 2DP
(0171–242 7583)

Glass Manufacturers Federation
19 Portland Place
London W1N 4BH
(0171–580 6952)

Heating and Ventilation Contractors
  Association
34 Palace Court
London W2 4JG
(0171–229 2488)

Institute of Plumbing
64 Station Lane
Hornchurch
Essex RN12 6NB
(017108 472791)

National Association of Scaffolding
  Contractors
18 Mansfield Street
London W1M 9FG
(0207–580 558)

National Council of Building Material
  Producers
26 Store Street
London WC1E 7BT
(0207–323 3770)

National Federation of Painting
  and Decorating Contractors
18 Mansfield Street
London W1M 9FG
(0171–580 5588)

National Federation of Plastering
  Contractors
82 New Cavendish Street
London W1M 8AD
(0171–580 5588)

National Federation of Roofing
  Contractors
24 Weymouth Street
London W1N 4LX
(0171–436 0387)

National Joint Council for Felt
  Roofing Contracting Industry
Fields House, Gower Road
Haywards Heath RH16 4PL
(01444 440027)

Royal Institute of British Architects
66 Portland Place
London W1N 4AD
(0171–580 5533)

Royal Institute of Chartered Surveyors
12 Great George Street
London SW1Y 5AG
(0171–222 7000)

MORTGAGE LENDERS

Build Store
Unit One Kingsthorn Park
Houston Industrial Estate
Livingstone EH54 5DB
(01506 417 130)

Britannia Building Society
Cheadle Road
Leek
Staffordshire ST13 5RG
(01538 399 399)

Buckinghamshire Building Society
High Street
Chalfont St Giles
Buckinghamshire HP8 4QB
(01494 879 500)

Newcastle Building Society
New Bridge Street
Newcastle Upon Tyne NE1 8AL
(0191 244 2000)

Skipton Building Society
The Bailey
Skipton
North Yorkshire BD23 1DN
(01756 705 000)

JOURNALS AND MAGAZINES

Build It
Isis Building
193 Marsh Wall
London E14 9SG
(020 777 8300)

INDEX

Abbreviations, x

Basement conversion, 31
Baths, 76
Bedroom extension, 31
Blockwork, 138–9
Boarded doors, 60
Brickwork, 138
Builders charges, xi-xii
Building Regulations, 15–17

Carpentry, 140
Casement windows, 62–4
Central heating, 31
Chimney pots, 38–9
Choosing a contractor, 5–6
Conservatories, 125
Contracts, 9–11
Copper pipework, 74
Cylinders, 74–5

Damp treatment, 100–1
Decorating, 84–91
Descriptions, ix–x
Disposal of material, 102–3
Doors:
  generally, 42–61
  boarded, 60
  external, 55–61
  fire doors, 49–50
  flush, 43–6, 47–50
  internal, 43–55
  garage, 60–1
  panelled, 46–7, 50–2, 55–60

  pine, 52–3
  wardrobe, 53–5
Double glazing, 31
Drainage, 144

Edgings, 93
Electrical work, 83–4
Equipment hire, 129–32
Estimates, 6
External doors, 55–61
External walls, 37–8
Extensions, 123–5
Extras, 8–9

Fencing, 94–5
Finance, 28–9
Fire doors, 49–50
Fireplaces, 35–7
Floor tiling, 70–1
Flush doors, 43–6, 47–50

Garage doors, 60–1
Garages, 31
Glazing, 79–83
Glossary, 145–7
Grants, 29–30

Heating, 77–9
Hours, materials and
  costs, 35–103
House extensions, 123–5

Insulation, 75
Internal doors, 43–55

Joinery, 140

Kitchen extension, 31
Kitchen fittings:
  generally, 66–9
  doors, 69
  drawer fronts, 68–9
  flat pack, 67
  ready assembled, 67–8

Loft conversion, 31, 125–6

Material costs, xi
Mensuration, 138
Metric system, 135–6
Mortar, 139–40

Ordering materials:
  generally, 22–6
  brickwork, 24–5
  concrete, 23–4
  excavation, 22–3
  floor tiling, 25
  painting, 25
  wallpapering, 26
  wall tiling, 25

Painting, 85, 142
Panelled doors, 46–7, 50–2,
  55–60
Paperhanging, 86, 143
Party walls, 22
Paths, 91–2
Patios, 95–8
Payments, 7–8
Pine doors, 52–3
Planning permissions, 17–22
Plastering, 70, 142
Plasterboard, 70
Plumbing:
  generally, 71–9
  alteration work, 72–4
  baths, 76
  copper pipework, 74
  cylinders, 74–5

  insulation, 75
  showers, 77
  sinks, 76
  storage tank, 74
  wash basin, 76
  WC suite, 76
Porch, 31
Programming, 26–8
Project costs, 123–6

Quotations, 6

Radiators, 77–9
Rising damp, 101–2
Roofing:
  generally, 38–42, 140–1
  flat, 42
  repairs, 39–41
  slates, 41
  tiles, 42
Rooflights, 66

Sash windows, 64–5
Self-building:
  generally, 107–19
  budgets, 115–18
  finance, 112–13
  plot finding, 108–12
  professional help, 113–15
Showers, 77
Sinks, 76
Skill level, xi
Slates, 41, 141
Storage tanks, 74
Swimming pools, 126

Temperature equivalents, 137
Tiling:
  floor, 70–1
  roof, 42
  wall, 71
Timber treatment, 99–100
Tool hire, 129–32

Useful addresses, 149–50
Using a Contractor, 3–11

Valves, 77
Variations, 8–9

Walling, 98–9
Wallpapering, 86, 143

Wall tiling, 71
Wash basins, 76
Wardrobe doors, 53–5
WC suites, 76
Windows:
  generally, 61–5
  casement, 62–4
  sash, 64–5

Milton Keynes UK
Ingram Content Group UK Ltd.
UKHW031151141024
449569UK00024B/878